MW00724054

Celebrating 125 Years of the U.S. Geological Survey

Compiled by Kathleen K. Gohn

Circular 1274

U.S. Department of the Interior
U.S. Geological Survey

U.S. Department of the Interior
Gale A. Norton, Secretary

U.S. Geological Survey
Charles G. Groat, Director

U.S. Geological Survey, Reston, Virginia: 2004

Free on application to U.S. Geological Survey, Information Services

Box 25286, Denver Federal Center
Denver, CO 80225

For more information about the USGS and its products:
Telephone: 1-888-ASK-USGS
World Wide Web: http://www.usgs.gov/

Suggested citation:
Gohn, Kathleen K., comp., 2004, Celebrating 125 years of the U.S. Geological Survey : U.S. Geological Survey Circular 1274, 56 p.

Library of Congress Cataloging-in-Publication Data

2001051109
ISBN 0-607-86197-5

Message from the Director

In the 125 years since its creation, the U.S. Geological Survey (USGS) has provided the science information needed to make vital decisions and safeguard society. In this anniversary year, we celebrate the mission that has guided us, the people and traditions that have shaped us, and the science and technology that will lead us into the future.

The early leaders of the USGS—King, Powell, Walcott—set high standards for the new organization, instilling a desire to achieve excellence in science and in service to the Nation. USGS accomplishments over the past 125 years have lived up to the vision laid out by the first Directors.

Today, the USGS continues to map, measure, and monitor our land and its resources and to conduct research that builds fundamental knowledge about the Earth, its resources, and its processes, contributing relevant and impartial information to critical societal issues.

As we celebrate our contributions and those made in partnership with the broader science community, we look with renewed energy to the coming decades. The USGS and the Nation will face many challenges during the coming years: water availability, climate change, habitat alteration, emerging diseases, invasive species, prediction of natural hazards before they become disasters, and many as-yet-unforeseen problems. We must be flexible and embrace new opportunities as they arise, to respond as new environmental challenges and concerns emerge and to seize new enhancements to information technology that make producing and presenting our science both easier and faster.

Through a wealth of long-term data and research, we have served the needs of society, the Earth, and its environment. This Circular captures a few of our past achievements, current research efforts, and hopes and challenges for the future. I invite you to celebrate our 125 years of service to the Nation as you read these pages and to join us as we look forward with a renewed sense of commitment to our mission of science for a changing world.

USGS Director Chip Groat.

Contents

Sidebars

Introduction

The USGS has been dedicated to providing credible, impartial science to inform critical decisions since its establishment by the Organic Act in 1879. Throughout its history, its functions have been expanded and altered in response to Congressional and Executive direction. The earliest activities of the USGS were geologic mapping and economic geology. Under John Wesley Powell, the second Director, topographic mapping and monitoring of the Nation's water resources became part of the USGS mandate.

Over the next century, parts of the USGS were incorporated into separate bureaus, including the Bureau of Reclamation, the Bureau of Mines, the Bureau of Land Management, and the Minerals Management Service. In the past decade, this trend has reversed, and parts of the Bureau of Mines and the National Biological Service have been incorporated into the USGS. The addition of biological research to geology, hydrology, and geography has strengthened the USGS by enabling an integrated approach to multidisciplinary science programs. The resulting synergy improves our ability to understand complex processes and systems and increases the value and applicability of USGS science.

This Circular presents the history of a wide range of scientific investigations in support of the Nation and shows how science has informed vital decisions about public health, public safety, public prosperity, and the use and conservation of natural resources. The topics highlighted here provide a glimpse of the rich legacy of research, current science efforts, and exciting opportunities for the future that will continue to provide solid dividends of progress in science and technology to the American public.

A Proud History of Service

Resource Information for a Strong Economy

One of the most critical needs recognized by the Organic Act of March 3, 1879, was for an understanding of the Nation's mineral and energy resources, as the economy moved from an agricultural to an industrial basis. The Federal Government began collecting and analyzing data and providing economic evaluation of mineral commodities in 1866. This effort was continued by the newly established USGS, which collected mineral statistics throughout the United States.

In addition, 1879 saw the start of comprehensive studies of three important mining districts—Leadville in Colorado and Comstock and Eureka in Nevada. In 1882, the USGS created a new division of mining statistics and technology to study the mineral resources of the United States. Studies focused on iron ore, coal, building stone, and petroleum—commodities that are still of interest today. Annual volumes describing the mineral resources of the United States were published beginning in 1882, and the series continues to this day. Studies of gold and coal in Alaska began in 1895; more than a century later, the USGS is still involved in assessing the resources of that vast region. In the late 19th century, developing industries in the United States were making growing demands on the Nation's natural resources, and assessments were increasingly needed. By 1900, geologists were beginning to develop models for how ores were formed.

Interest in non-metal resources, including fossil fuels,

Corbis Images

1849

On March 3, Congress establishes the Department of the Interior (DOI).

1879

The Organic Act of March 3 establishes the U.S. Geological Survey to classify the public lands and examine the geological structure, mineral resources, and products within and outside the national domain.

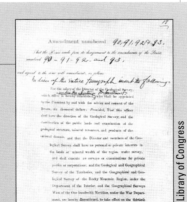
Library of Congress

1879–1881

Clarence King is the 1st Director of the USGS.

John Wesley Powell

In the spring of 1869, a 35-year-old, one-armed, Civil War veteran led an expedition down the Colorado River into a great, unknown, uncharted territory. Ninety-nine days later, after one of the most daring journeys in American history, John Wesley Powell emerged from the Grand Canyon to become a contemporary American hero.

Powell was born in 1834 at Mount Morris in New York. He sporadically attended Wheaton and Oberlin Colleges but never earned a degree. Interested in botany and geology at an early age, he began his scientific investigations with a series of self-directed field trips, including a rowboat voyage that covered the length of the Mississippi River. In 1861, Powell enlisted in the Union Army and was commissioned as captain. He lost his right arm at the elbow in the Battle of Shiloh but returned to active duty and was promoted to the rank of major.

Powell led geological and ethnological explorations in Arizona and Utah under the auspices of the Smithsonian Institution. His efforts toward reorganizing early surveys in the West helped establish the USGS in 1879. In March 1881, Powell became Director of the USGS when the first Director, Clarence King, resigned. Powell championed a nationwide program of topographic mapping, promoted systematic studies and data collection in hydrography, and advocated conservation and careful planning in the use of Western lands. His cautionary views regarding settlement in the West antagonized influential Western politicians, and Powell resigned from the USGS in 1894.

Powell was widely recognized as one of the leading scientists of his age. He was a founder and president of the Cosmos Club in Washington, a founder and president of the Anthropological Society of Washington, one of the earliest members of the Biological Society of Washington, and an organizer of the Geological Society of Washington. He helped establish the National Geographic Society and the Geological Society of America. In 1888, he was elected president of the American Association for the Advancement of Science, then considered the highest honor for an American scientist, and he received honorary degrees from several universities at home and abroad. Throughout his career, in combination with his work in geology, geography, and botany, Powell studied the West's Native Americans and their languages. He founded and, in 1879, was named the first director of the Smithsonian Institution's Bureau of Ethnology, a position he held until his death in 1902.

increased during the early 20th century; by the mid-1920s, so much oil and gas had been located in the Gulf Coast, mid-continent, and California that there was a gas surplus. The two World Wars and the Korean War focused USGS and national attention on strategic minerals such as tin, nickel, platinum, nitrates, and potash that were needed for the war effort; the search for new resources expanded to Central and South America. After World War II, exploration for fossil fuels and uranium increased dramatically; an area of particular interest was the 23 million acres of Alaska's North Slope in Naval Petroleum Reserve No. 4—now better known as the National Petroleum Reserve–Alaska.

In 1964, Congress passed the Wilderness Act setting more than 9 million acres of national forest lands aside from permanent roads, buildings, and commercial activities. The USGS and the U.S. Bureau of Mines were asked to assess the mineral resources in each area of the proposed or established wilderness by the end of 1983, after which time no new mining claims would be allowed. The amount of land to be assessed kept growing, finally reaching 45 million acres that included Bureau of Land Management and other Federal lands in addition to the national forest areas. To complete the assessments by the due date of 1983, USGS scientists developed tools and techniques that combined geology, geochemistry, geophysics, and computer modeling to produce estimates of the potential for undiscovered mineral resources on the wilderness lands. These tools and techniques continue to be refined by scientists working with Federal partners such as the Forest Service and other DOI bureaus to provide science for informed decisions.

1879

The first USGS National Center is established at 803 G St., NW., in DOI-rented space in the Patent Office Building. King establishes regional centers in Denver, Salt Lake City, and San Francisco. King also establishes the Mining and General Geology Divisions.

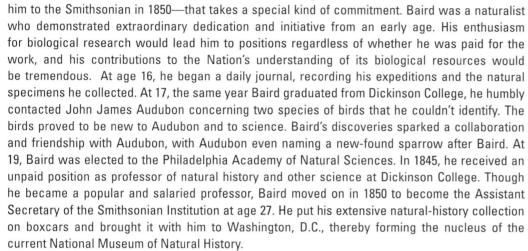

Smithsonian Institution

Spencer F. Baird

While many of us are talented collectors and enthusiastic pack rats, it's doubtful that any of us will rival the 89,000 pounds of natural history specimens that Spencer F. Baird brought with him to the Smithsonian in 1850—that takes a special kind of commitment. Baird was a naturalist who demonstrated extraordinary dedication and initiative from an early age. His enthusiasm for biological research would lead him to positions regardless of whether he was paid for the work, and his contributions to the Nation's understanding of its biological resources would be tremendous. At age 16, he began a daily journal, recording his expeditions and the natural specimens he collected. At 17, the same year Baird graduated from Dickinson College, he humbly contacted John James Audubon concerning two species of birds that he couldn't identify. The birds proved to be new to Audubon and to science. Baird's discoveries sparked a collaboration and friendship with Audubon, with Audubon even naming a new-found sparrow after Baird. At 19, Baird was elected to the Philadelphia Academy of Natural Sciences. In 1845, he received an unpaid position as professor of natural history and other science at Dickinson College. Though he became a popular and salaried professor, Baird moved on in 1850 to become the Assistant Secretary of the Smithsonian Institution at age 27. He put his extensive natural-history collection on boxcars and brought it with him to Washington, D.C., thereby forming the nucleus of the current National Museum of Natural History.

While serving at the Smithsonian, Baird undertook fish studies at Woods Hole, Mass. In 1870, the coastal fisheries in southern New England were declining, and there was concern over the dwindling food source. Baird saw this as the perfect opportunity to serve the Nation by tackling the problem with science. In 1871, he persuaded Congress to create the U.S. Commission of Fish and Fisheries, where he served as Commissioner without additional pay until his death in 1887.

1879

The mining geology program begins with (1) the collection of mineral statistics in both the Eastern and Western States and (2) comprehensive studies of the geology and technology of three great mining districts—Leadville in Colorado and Comstock and Eureka in Nevada.

1880

The first chemistry laboratory is established in the USGS, and the first geophysical work begins as part of the study of the Comstock mining district.

C. Hart Merriam

Like many naturalists, C. Hart Merriam began his studies as a boy exploring the woods around his home. He joined the Hayden Survey at age 16 and traveled to Yellowstone and the Western Territories. At 18, he published a 50-page scientific report of his studies. Though he spent several years practicing medicine, Merriam returned to his biological and ecological roots, where his work was controversial, influential, and essential to the foundation of what is now the USGS Biological Resources discipline.

He was a charter member of the American Ornithologists' Union (AOU), where he served as chairman of the Committee on the Migration of Birds. In 1885, he was appointed to head what was then the Section of Economic Ornithology with the U.S. Department of Agriculture. Merriam's wide interests led to the section's expansion into the Division of Economic Ornithology and Mammalogy, which was later renamed the Division of Biological Survey.

Library of Congress, Ruthven Deane Collection

Merriam's scientific work has had long-lasting effects. Research on bird distribution and migration that he started at AOU continued at the Biological Survey and developed into a network of volunteer observers spanning the United States. The program served as the foundation for modern surveys that support national and international bird conservation efforts. Perhaps Merriam's most noted work is his delineation of life zones for North America. These life zones, originally drawn on USGS base maps, furthered ecological understanding and continue to influence biologists and ecologists even today.

Florence Bascom

Florence Bascom collected many "firsts" in her geological career: she was the first woman to receive a Ph.D. from Johns Hopkins University (sitting behind a screen so the male students wouldn't know she was there), the first woman geologist hired by the USGS, the first woman to present a scientific paper at the Geological Society of Washington, and the first woman officer of the Geological Society of America.

Bascom was born in 1862 in Williamstown, Mass., and died in Northampton 83 years later. Her father, president first of Williams College and later of the University of Wisconsin, encouraged her interest in geology, and she earned bachelor's and master's degrees in geology from the University of Wisconsin in the 1880s. Her professors at Wisconsin, Roland Irving and Charles Van Hise, were also employed by the USGS, as was her professor at Johns Hopkins, George Williams. After receiving her Ph.D. in 1893, she began teaching geology at Bryn Mawr College in 1895, but she combined her teaching career with active field and laboratory work for the USGS. Bascom retired from teaching in 1928 but continued to work for the USGS until 1936.

She was an authority on the rocks of the Piedmont and published maps and folios; she also studied water resources of the Philadelphia region. Her writing was vigorous and incisive; her conversation was forceful and clear, if sometimes caustic. She developed the geology curriculum at Bryn Mawr from a single course to a full major and then a graduate program, which trained most American women geologists during the first third of the 20th century. At least three of her students later joined the USGS: Eleanora Bliss Knopf, Anna Jonas Stose, and Julia Gardner.

The analytical techniques and deposit models that were developed for assessing wilderness lands enabled USGS scientists to conduct a national mineral assessment in 1998 estimating that the United States still had as much undiscovered gold, silver, copper, lead, and zinc as had already been identified to date; a national oil and gas assessment in 1995 and a world petroleum assessment in 2000 showed that more undiscovered energy resources were left than had been used to date. The techniques, analytical procedures, and models that have been developed, and the resulting assessments of national and global resources, are valued by domestic and international customers and stakeholders. USGS scientists lead the world in the ability to conduct mineral and energy resource assessments, providing essential information for sound energy policy, environmental decisions, and land and resource management.

Biological Reconnaissance and Research

In the early years of the Nation, little was known about the species of plants and animals that occurred in the United States and their distributions across the landscape. The Lewis and Clark expedition in 1804–1806 made exceptional breakthroughs, but it was not until after the Civil War that the Federal Government established agencies with responsibility for scientific research and exploration. Naturalists Spencer F. Baird and Clinton Hart Merriam were pioneers who laid the foundations for research that continues in the USGS today. While serving as Assistant Secretary of the Smithsonian, Baird became concerned about declining fisheries in New England and persuaded Congress to create the U.S. Commission of Fish and Fisheries in 1871; Baird was selected to lead the commission. In 1886, the Division of Economic Ornithology and Mammalogy was established in the Department of Agriculture, with Merriam appointed as its first Chief. Much of the Division's

USFWS

1881

The second USGS National Center is established in the National Museum Building (now the Arts and Industries Building of the Smithsonian Institution).

1881–1894

John Wesley Powell is the 2d Director of the USGS. Powell looks on geology and topography as independent, although closely related, parts of the greater field of geography and makes the topographic work of the USGS independent of geologic studies.

Smithsonian Institution

early work focused on studying the value of birds to control agricultural pests and defining the geographical distribution of animals and plants throughout the country.

By 1906, this Division was transformed into the U.S. Bureau of Biological Survey. Much of the discovery of new species of terrestrial wildlife and mapping of basic vegetation types in the United States at the turn of the century was accomplished by this group, particularly in the West. The Commission of Fish and Fisheries was combined with the Bureau of Biological Survey in 1940 to form the Fish and Wildlife Service (FWS) in the Department of the Interior, institutionalizing the relationship that Baird and Merriam had begun as fellow naturalists 60 years earlier.

Research on the biogeography, taxonomy, and ecological relationships of fish, wildlife, and plants continued in FWS and is maintained today with the reassignment of these programs to the USGS in 1996. Much of the research on the occurrence, population dynamics, ecosystem functioning, and remote sensing of biological resources currently undertaken by USGS, especially on Department of the Interior lands, is a continuation of the early days of biological survey and reconnaissance of the U.S. frontier lands. The USGS still has responsibility for curating the North American mammal, bird, reptile, and amphibian collections at the Smithsonian, including nearly 5,000 specimens donated by Baird in 1850.

Geologic and Topographic Mapping

Maps play a fundamental role in conveying the breadth of our national domain, in building our understanding of natural resources and earth history, and in providing place-based information required for Government administration, water issues, mining, transportation, agriculture, and forestry. As the Nation grew westward, many national leaders recognized that accurate maps of the landscape and the rocks beneath the surface were essential building blocks for economic development. When the USGS was established in 1879, it was assigned the task of "...classification of the public lands and examination of the geological structure, mineral resources and products of the national domain..." (Organic Act of the U.S. Geological Survey, U.S. Statues at Large, v. 20, p. 394). The Organic Act of 1879 specifically mentions geologic and economic maps as among the expected products of these investigations. In 1882, at the instigation of Director J.W. Powell, the USGS established a national program for systematic

Henry Gannett

If you have ever used a topographic map to find your way around some remote part of the United States, or even been confused by place names of populated areas, you'll appreciate the pioneering work of Henry Gannett, an early American geographer often considered to be the father of American topographic mapping.

He began his career in topographic mapping with the Hayden Survey in 1871 and recognized early on the importance of geography as the cornerstone for other sciences. He worked zealously to present geographic knowledge so that it could be widely utilized by diverse audiences. Many enduring methods and standards of USGS mapmaking were developed under his leadership.

The USGS Geography Program was established under his direction, and he served as Chief Geographer from 1882 to 1914. Under his command, the program's first topographic map sheets were produced and the program became the Division of Geography.

Through his work as geographer of the U.S. censuses of 1880, 1890, and 1900 and the Philippine, Cuban, and Puerto Rican censuses, he became interested in place names. His efforts to resolve difficulties caused by the confusion of names contributed to the establishment of the U.S. Board on Geographic Names in 1890. He served as the Board's chairman from 1894 until 1910.

Gannett was also one of the founders of the National Geographic Society (president, 1910–14), the Geological Society of America, and the Association of American Geographers.

1882
The USGS Library, authorized in 1879, is formally established with a collection of 1,400 books and Charles C. Darwin as Librarian.

1884
The third USGS National Center is established at 1330 F St., NW.

Grove Karl Gilbert

Grove Karl Gilbert, the first Chief Geologist of the USGS, was one of the founders of modern geomorphology, the study of landforms. He recognized that landforms reflect a state of balance between the processes that act upon them and their structure and composition. Before the USGS was established, Gilbert worked in Utah under George Wheeler, who led a survey to map the United States west of the 100th meridian, during 1871–73, and John Wesley Powell. In 1900, Gilbert won the Wollaston Medal, the Geological Society of London's most prestigious award. He was only the third American to be honored with the award.

His last great geological contribution came in 1905, when he was sent to California in connection with hydraulic gold mining. He studied the fluvial character of the Sacramento River and the impact of mining debris, conducted the first major quantitative modeling of streamflow in large flumes, and observed the effects of the great 1906 earthquake in San Francisco. His reports on hydraulic mining were masterpieces of engineering geology.

Gilbert is considered one of the most distinguished American geologists and was described as a "great engine of research" by author Stephen J. Pyne. He made significant contributions to the fields of tectonics, hydrology, glaciology, earthquake studies, and geological methods.

topographic mapping that has provided a sound foundation of accurate, widely available geographic information for officials at all levels of government, as well as for scholars, students, and the public. Powell spoke passionately of his vision for comprehensive topographic maps for the Nation in testimony before Congress in 1884, "A Government cannot do any scientific work of more value to the people at large, than by causing the construction of proper topographic maps of the country" (Sundry Civil Expenses Act, August 7, 1882, 22 Stat.L., p. 329).

Also in 1882, Congress authorized the USGS to continue preparation of the geological map of the United States, which Powell interpreted as authorizing extension of the Survey's geologic mapping throughout the Nation. The first step was the establishment of the Geologic Atlas of the United States, to consist of a series of folios in a standard format, each containing topographic, geologic, and other maps and illustrations and text describing the geology of a particular quadrangle. The first of these folios was published in 1894; in the next decade, 106 were published. The series came to an end when it became clear that folios were too generalized, and attention shifted to more focused geologic mapping to address issues such as the petroleum potential of the North Slope of Alaska or uranium resources of the Colorado Plateau.

The Geologic Mapping Act of 1992 recognized the importance of geologic mapping for the utilization of mineral resources, effective stewardship of the environment,

---

1885
The Division of Economic Ornithology and Mammalogy is established in the Department of Agriculture (USDA) with C. Hart Merriman as the first Chief.

1886
Charleston, S. C., earthquake—magnitude 7.3.

1889
The USDA Division of Economic Ornithology begins the North American Fauna series and publishes Merriam's concept of life zones of North America.

1889
Powell assigns Frederick Haynes Newell to lead a team to develop and standardize technology for measuring streamflow in rivers. The work takes place at a camp on the Rio Grande River near Embudo, N. Mex., and is the location of the USGS's first streamgage.

safe disposal of domestic and industrial waste, and mitigation of natural hazards. It established a program under the leadership of the USGS, with advice and contributions from other Federal and State agencies, academia, and the private sector, to help meet these needs. Currently, much USGS geologic mapping focuses on studies of immediate practical concern such as the distribution of active faults that pose earthquake threats to population centers or critical installations, materials that are particularly subject to landslides or slope failure, or volcanic deposits that help delineate areas at risk from volcanic eruptions; studies of the movement of water, petroleum, natural gas, and waste (including radioactive waste) through the Earth; and studies of surficial deposits that record climatic changes over time. All of these can be addressed through use of geologic maps.

Powell's vision for complete topographic coverage of the Nation persisted. The present 7.5-minute series of USGS topographic maps, mostly at 1:24,000 scale, was begun in the 1930s and completed in 1991 for all of the United States except Alaska, which is largely mapped at 1:63,360 scale. Thirty-three million person hours and $1.6 billion were invested in the complex endeavor to depict the physical and cultural nature of our land as a resource to all Americans. Portrayed on 55,000 separate map sheets, the 7.5-minute series of large-scale USGS topographic maps remains the most comprehensive coverage of the Nation's landscape and its critical infrastructure.

Mapping is alive and well in the USGS. Today's maps rely on satellite imagery, aerial photography, Global Positioning System (GPS) technology, and geographic information system

Frederick Haynes Newell

Bureau of Reclamation

"He carried water from a mountain wilderness to turn the waste places of the desert into homes for freemen." This is the inscription on the Cullum Geographical Medal that F.H. Newell received in 1918 for his contributions to the reclamation of the arid West. Newell was the second Commissioner of what would become the Bureau of Reclamation; his legacy is important not only to the irrigation of the West but also to the creation of the Nation's streamflow monitoring network.

Newell began his work as an assistant hydraulic engineer for Major John Wesley Powell, second Director of the USGS. Powell foresaw the value of irrigation to the future economy of the Western United States, but in 1888 the techniques and instruments needed to measure streamflow had not been developed. Newell created a training camp at Embudo, New Mex., to develop standard methods of measuring streamflow and to train the people to do it.

USGS now has more than 7,000 streamgages nationwide, and the techniques Newell developed have been used to measure and monitor streamflow in the United States for more than a century. These gages record the highs and lows of an essential resource, mark disastrous floods, and provide the accurate and objective information needed to manage water and protect communities.

President Theodore Roosevelt cited him as an advisor and counselor in the development of his strategy for irrigation and referred in his autobiography to Newell's devotion to the establishment of the Bureau of Reclamation and his leadership, high character, and "constructive imagination."

1894–1907
Charles Doolittle Walcott is the 3d Director of the USGS. Walcott abolishes all organizational units within the Geologic Branch and assumes direct control of the work. Mining geology studies are resumed and extended into the Eastern States, and a study of gold deposits, including exploration for new sources, is begun. The deposits at Cripple Creek, Colo., and at Mercur, Utah, are usable through development of the cyanidation process.

1894
A small appropriation is obtained for the purpose of "gauging the streams and determining the water supply of the United States."

1895
Basic science is an integral part of the Geologic Branch program. Fundamental studies are made in the genesis of ore deposits, in paleontology and stratigraphy, in glacial geology, and in petrography. The geologic time scale is revised, and new definitions for rock classes are developed.

Luna Leopold

During his 10 years as USGS Chief Hydrologist, Luna B. Leopold transformed the USGS Water Resources Division into the Nation's premier agency for water research. From the outset, his work was creative, pioneering, and multidisciplinary.

Leopold's greatest impact on the earth sciences began with the 1953 publication of USGS Professional Paper 252, "The Hydraulic Geometry of Stream Channels," a paper that provided a basis for observing rivers throughout the world. With this paper, Leopold initiated a new era in the study of rivers, one that involved quantitative approaches that spread to the broader field of geomorphology, the study of the evolution and configuration of landforms. His research consistently related meteorology and climatology to landscape process, a concept that has become a central feature of geomorphology.

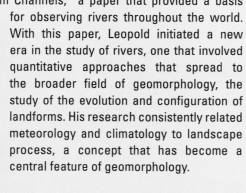

A son of Aldo Leopold, one of the early leaders of the movement to preserve the American wilderness, Luna Leopold was a pioneer in making ecological foresight a part of hydrological studies and of the information provided for those making water-resource policies. In USGS Circular 620, published in 1969, Leopold coined the term "environmental impact." He helped create a framework for environmental impact statements, giving direction to the growing trend for decisionmakers to consider environmental impact before starting a project.

Fortunately for future generations, Leopold's enthusiasm for rivers has proved contagious. He has mentored many prominent scientists and shared his creative intellect and passionate commitment to science and society, inspiring generations of colleagues and students to use their talents both to the pursuit of science and to its application to society.

(GIS)-based cartography. In the end, however, geologic mapping comes down to a geologist walking the ground with compass, pick, notebook, and hand lens, just as in Powell's day.

Measuring the Nation's Streamflow

Information to define, use, and manage water resources was critical to the settling of the West, and water continues to be a major issue for the Nation and the world. In 1889, John Wesley Powell assigned Frederick Haynes Newell to lead a group of recent engineering graduates and other young men to develop and standardize a technology for measuring streamflow in rivers. The men established the first streamgaging station operated in the United States by the USGS on the Rio Grande near Embudo, N. Mex., as part of the effort to train individuals to

measure the flow of rivers and streams and to define standard streamgaging procedures As soon as methods were developed at Embudo, USGS scientists and engineers were sent throughout the West to collect streamflow data.

By 1891, the first streamflow measurement in the East was made on the Potomac River at Chain Bridge, near Washington, D.C. In 1895, the first Cooperative Program in the Nation began in Kansas, through an agreement with the newly established Kansas Board of Irrigation Survey and Experiment (now the Division of Water Resources of the Kansas Department of Agriculture). This agreement provided for measurement of streamflow at seven sites to ascertain water-supply potential. By 1900, only 163 stations were in operation. Most of the stations were in the West and were used to satisfy needs for

irrigation. As concerns grew about floods, droughts, and increased use of water for irrigation and hydroelectric power, the USGS streamgaging program expanded to fill the Nation's needs. Congress passed legislation in 1929 that officially recognized the Cooperative Program, in which costs are shared with State and local agencies, and in the ensuing years, cooperative streamgaging programs were established with many State and local agencies. Today, more than 90 percent of USGS streamgages are operated in partnership with other Federal, State, and local agencies.

1896
Benchmarks and map sales are authorized.

1898
The U.S. Senate calls for a Division of Mines and Mining in the USGS to gather statistics on mineral resources and mineral production and to make investigations related to mines and mining.

1904
In 1904, the USGS celebrates its 25th anniversary, having come from an organization of 38 employees at the end of its first year to one of 491 employees (and another 187 in the adjunct Reclamation Service). Its appropriation is $1.4 million. It is the leading geologic institution in the United States in the view of the publication "American Men of Science," and many of the 100 geologists whose work is considered most significant by their peers, including the first five ranked numerically, are associated with the USGS.

USDA, U.S. Forest Service

Severe drought in the early 1930s and the floods in 1936–37 in the Ohio and the Potomac River basins increased the awareness among Federal, State, and local agencies that managing the Nation's water resources requires comprehensive, reliable streamflow data. By 1950, more than 6,000 streamgages were in use across the landscape. The data collected at these gages are used for forecasting floods, scheduling hydropower production, characterizing water quality, and even helping a family decide whether it's a good time to go fishing or boating. Although the USGS currently uses many new technical advancements to measure streamflow and operate gaging stations, procedures developed by Newell and his colleagues at Embudo still form the basis for the nationwide streamgaging program carried out to this day by the USGS.

Natural Hazards

Floods, volcanic eruptions, earthquakes, landslides, drought—these natural hazards have reshaped the landscape for millions of years and have been the focus of USGS research since its beginnings. The USGS has Federal responsibility for issuing warnings on geologic hazards—to get people out of harm's way—as well as providing information on how to live and build safely—to keep people and property out of harm's way.

Early reports by USGS scientists described Hawaiian volcanoes and the great Charleston, S.C., earthquake of 1886. The San Francisco earthquake of 1906 was one of the most significant earthquakes of all time—not just because of its devastating impact on the city but also because of the wealth of scientific knowledge derived from it. Rupturing more than

250 miles of the San Andreas fault from northwest of San Juan Bautista to Cape Mendocino, Calif., the earthquake confounded contemporary geologists.

The significance of the fault and recognition of its large cumulative offset would not be fully understood until the emergence of plate tectonic theory more than half a century later. Developments in plate tectonics during the 1960s gave scientists a revolutionary new way of understanding the Earth and its processes. Basic research by USGS scientists at Menlo Park, Calif., on the Earth's magnetic field provided essential supporting information for the

1905
The USDA Division of Biological Survey becomes the Bureau of Biological Survey.

1906
San Francisco, Calif., earthquake— magnitude 7.8.

1907
The Reclamation Service becomes an independent agency. F.H. Newell leaves the USGS to become the Service's Director.

1907–1930
George Otis Smith is the 4th Director of the USGS. Smith believes that the work of the USGS should be primarily "practical."

David A. Johnston

David Johnston was a 30-year-old volcanologist with the USGS when he was swept away by the catastrophic eruption of Mount St. Helens on the morning of May 18, 1980.

Johnston came to the USGS in 1978 to work on volcanic gases. He expanded the program for monitoring volcanic emissions in Alaska and the Cascade Range, with the goal of determining whether or not changes in gas geochemistry might provide warning of impending eruptive activity. It was therefore natural that Johnston was one of the first geologists on the volcano when Mount St. Helens reawakened in March 1980.

As one of the first members of the USGS monitoring team to arrive at Mount St. Helens and the scientist in charge of volcanic-gas studies, Johnston spent long hours working on and close to the mountain. From his experience with active Alaskan volcanoes, Johnston understood better than most the hazards of explosive volcanism. At the same time, he repeatedly voiced the conviction that adequate hazard assessments require accepting the dangers of onsite monitoring of active volcanic processes. The volcano-monitoring effort that Johnston was a part of helped persuade the authorities first to limit access to the area around the volcano and then to resist heavy pressure to reopen it, thereby holding the May 18 death toll to a few tens instead of thousands.

Johnston was an exemplary scientist, and his approach to his work was a model for all: dedicated and hard-working, with meticulous organization and observation followed by careful evaluation and interpretation. At the same time, Johnston was clearly genuine, with an infectious curiosity and enthusiasm. But perhaps his most essential quality was the ability to look for, see, and encourage the best in everyone.

Lucille Stickel

In 1946, when Lucille Stickel published her first contaminant paper, a study of the new pesticide DDT, virtually nothing was known about the harmful effects of pesticides on wildlife. The impact of her pioneering research would be significant and far reaching, as Rachel Carson used Stickel's research as the basis for much of her book, Silent Spring. The book created a public understanding of the importance of preserving the Earth's resources and ushered in a new age of environmental awareness.

Stickel began her career as a Junior Biologist in 1943 at the Patuxent Wildlife Center. She became one of the early pioneers in the fledgling field of wildlife toxicology and is responsible for many historic findings and techniques. From 1973 to 1981, she served as Director of the Center, making her the first woman to head a major Fish and Wildlife Service laboratory. In 1989, the Chemistry Building at the Patuxent Center was renamed Stickel Laboratory in appreciation of the decades of dedicated service by Stickel and her husband Bill, her lifelong research partner and collaborator.

Stickel's commitment to the preservation of wildlife and natural resources was recognized in 1998 when the Society of Environmental Toxicology and Chemistry honored her with its prestigious Rachel Carson Award. She also received the DOI's Distinguished Service Award and the Aldo Leopold Award of the Wildlife Society.

new theory, which explains the causes and distribution of earthquake and volcanic zones around the world. For the past 25 years, USGS scientists have been leading the research effort to understand the causes and effects of earthquakes and to apply that knowledge to reduce earthquake risk across the Nation. USGS national seismic hazard maps form the foundation for national building codes and are making billions of dollars of new construction safer from earthquakes each year.

The United States has more than 50 historically active volcanoes within its borders, more than any other country except for Indonesia and Japan. The Aleutian volcanic arc in Alaska produces an average of 1 to 2 eruptions per year. Hawaii's most active volcano, Kilauea, has been erupting almost continuously since January 1983, making the current eruption its longest in more than 600 years. The Cascade Range in the Pacific Northwest produces fewer eruptions, an average of 1 to 2 per century, but threatens more residents than the volcanoes of Alaska and Hawaii combined. The devastating landslide and eruption of Mount St. Helens, Wash., on May 18, 1980, was the impetus for creation of the David A. Johnston Cascades Volcano Observatory, which bears the name of the USGS geologist who died in the eruption at a forward observation post on what is now called Johnston Ridge. The Mount St. Helens eruption was a turning point for modern volcanology that led to numerous discoveries and improved eruption prediction techniques.

Protecting Biodiversity—Research on Endangered Species

By the late 1800s, the wide-scale clearing of land for agriculture and intensive hunting of wildlife for commercial markets had taken their toll on many native species. Of the millions of bison that had roamed the North American plains, fewer than 1,000 remained. Passenger pigeons, whose immense flocks had once darkened the skies, were nearing extinction. Populations of snowy egret and other colonial-nesting wading birds had been reduced to a small fraction of their historical size. Alarmed by the rapid loss of our natural heritage, President Theodore Roosevelt created the first Federal Bird Reservation on Pelican Island, Fla., in 1903. Congress passed several laws in the early part of the 20th century to protect

1910

The Bureau of Mines is founded. The USGS Technologic Branch is transferred to the Bureau of Mines, and structural materials testing is transferred first to the Bureau of Mines, then to the Bureau of Standards. Joseph A. Holmes, head of the USGS's Technologic Branch, is the first Director of the Bureau.

1912

The Hawaiian Volcano Observatory is established. Under the directorship of Massachusetts Institute of Technology professor Thomas A. Jaggar (1912-1940), HVO scientists conduct pioneering studies of volcanic processes.

USFWS

wildlife populations, the most notable being the Migratory Bird Treaty Act (1918), which established Federal jurisdiction in the hunting of migratory birds.

The Patuxent Wildlife Research Center (then part of the U.S. Department of Agriculture), in Patuxent, Md., began research on migratory waterfowl in the mid-1930s. The whooping crane was one of the first species studied, and the effort to conserve this majestic bird later came to epitomize the endangered species movement. World War II interrupted the research, but after the war scientific efforts at Patuxent began anew, focusing on trumpeter swans and later on rapidly declining bald eagle populations.

By the time the Endangered Species Act was passed in 1973, a considerable amount of research associated with protecting imperiled species was

already underway, including work on black-footed ferrets, wolves, Puerto Rican parrots, California condors, and several Hawaiian birds. The approach was two-pronged: (1) to study species in their native habitat to identify factors affecting distribution and abundance and (2) to conduct research on species in captivity to develop captive breeding techniques for releasing individuals to the wild to bolster wild populations. Of the nine captive populations studied at Patuxent, two (the Aleutian Canada goose and the bald eagle) have been removed from the endangered species list, and five more (the whooping crane, timber wolf, masked bobwhite,

California condor, and Mississippi sandhill crane) are on the road to recovery. Research on species such as the gray wolf and American alligator has enabled their populations to recover or stabilize.

Over the past 30 years, research on endangered species has expanded to include a wide range of investigations at nearly every USGS Science Center and Cooperative Research Unit across the United States. USGS scientists are examining endangered species' life history characteristics, habitat requirements, population dynamics, migration patterns, and genetic relationships. They are evaluating the success of conservation actions, advising recovery teams, and working side by side with natural resource managers in the Fish and Wildlife Service, National Park Service, Bureau of Land Management, and other DOI bureaus.

Patuxent Wildlife Research Center

The establishment and evolution of the Patuxent Wildlife Research Center mark important events not only in the history of the USGS but in the changing attitudes and policies of the Nation. They signify a broadening of concern from wildlife's impact on people to an equal emphasis on people's impact on wildlife.

In the early part of the century, most wildlife research focused on the negative effects of wildlife on agriculture. The Dust Bowl era in the 1930s created a public desire to help restore wildlife populations, and the Patuxent Research Refuge was established in 1936 as the Nation's first wildlife experiment station.

The successes of the Center have been numerous. The captive propagation program attained international prominence by helping to increase the wild populations of bald eagles, whooping cranes, and other species throughout North America; urban wildlife research established development-planning techniques used throughout the country; and advice on backyard bird feeding is used by millions of homeowners every year. Perhaps the Center's best known success came in 1969, when researchers linked eggshell thinning with DDT. This discovery and testimony of Patuxent researchers before Congress resulted in the nationwide banning of DDT and other organochlorine pesticides in 1972.

Today, research at Patuxent continues to address national wildlife concerns and to maintain partnerships with State and municipal governments concerning local wildlife research issues. The continuation of this long relationship with our partners, including FWS and the National Park Service, will help ensure that USGS science informs decisionmakers charged with protecting and managing our Nation's living resources.

1916

The USGS reorients its work to aid the search for both metals and fuels, extending the search to Central and South America and the West Indies.

1917

The United States enters World War I in April, and the USGS forms a Division of Military Surveys. The strategic minerals concept is born. During these years, the USGS is the main source of information on mineral production.

1917

The fourth USGS National Center is established in the Interior Building on 18th & E Sts., NW.

1922

The USGS becomes involved in energy policy. A Coal Commission is established, and USGS resource data provide a basis for the Commission's report.

Denver Federal Center

For 58 years, the USGS has worked in a center that was once a Federal munitions plant. In December 1940, Denver, Colo., was selected as the site of one of several plants that would be part of future U.S. war efforts. The Denver Ordnance Plant, consisting of more than 200 buildings, was dedicated on October 25, 1941. In support of the war, it produced .30-caliber ammunition, heavy artillery shells, and 8-inch and 144-millimeter shells.

After Japan surrendered on August 15, 1945, the plant's production days were numbered. By October, the Denver Ordnance Plant had been declared surplus property. It was turned over to the Reconstruction Finance Corporation, and the decision was made to use the site to implement a 1938 plan to develop a Federal presence in Denver. The Denver Ordnance Plant became the Denver Federal Center (DFC) and is now home to 21 agencies and about 8,500 people.

The USGS began its occupancy of the DFC in July 1946 in Building 25. It wasn't long before Buildings 5, 21, 82, and parts of 51 and 20 were also occupied by USGS, as the DFC became home to the Topographic, Geologic, Water Resources, Conservation, Publications, and Administrative Divisions.

In 1976, a fire destroyed about 22,000 square feet of laboratories, but that didn't prevent the USGS from expanding its presence on the premises. Today, there are nearly 1,400 USGS personnel, contractors, and emeriti working on the DFC, where the USGS uses just over 1.4 million square feet of space in 18 buildings.

The Endangered Species Act now protects 1,265 species in the United States, including 519 species of animals and 746 species of plants. Hundreds more are listed as candidate species. The need for scientific tools to protect such a diverse group of organisms and their habitats is immense. However challenging the task, USGS scientists are hard at work developing practical solutions to the many problems that confront America's endangered species.

North American Breeding Bird Survey

In the mid-20th century, the success of DDT as a pesticide ushered in a new era of synthetic chemical pest control. As pesticide use grew, concerns about the effects on wildlife began to emerge. Local studies had attributed some bird kills to pesticides, but it was unclear how, or if, bird populations were being affected at regional or national levels. In 1963, responding to this concern, Chandler Robbins and colleagues at the Patuxent Wildlife Research Center began developing the North American Breeding Bird Survey (BBS) to monitor bird populations over large geographic areas. The BBS, jointly coordinated by the USGS and the Canadian Wildlife Service, is the foundation of modern non-game, land-bird conservation in North America. BBS data can be used to estimate trends in bird populations; declining numbers send an early warning signal that galvanizes

research and management action to determine the cause of population declines and reverse them before it's too late. Data are used by the U.S. Fish and Wildlife Service, Canadian Wildlife Service, Partners in Flight, and State agencies to monitor bird species, determine which species might be declining, and develop conservation goals.

Skilled volunteer birders spend one or two mornings a year at the height of the avian breeding season to collect population data along roadside survey routes. The collection protocols are carefully established to ensure that data can be reliably compared year by year. There are more than 4,100 BBS routes scattered across the United States and Canada, of which about 3,000 are sampled annually. More than 2,200 birders participate each year; more than 8,400 people have participated during the

1925

DOI delegates the responsibility for supervising mineral lease operations on the public lands to the USGS. This mandate requires a large force of mining and petroleum engineers, increasing USGS staff to more than 1,000 employees, only 126 of whom are geologists.

1929

The USGS is 50 years old. Its appropriation is $2 million, with total funds of $3.4 million. There are 998 permanent employees who conduct mapping and research investigations in 45 States, Alaska, Hawaii, and the District of Columbia. Nearly 44 percent of the continental United States (excluding Alaska) has been topographically mapped. Streamflow is being measured at 2,238 gaging stations, and the income from mineral leases, licenses, and prospecting permits on the public lands under USGS supervision is $4.1 million.

★ 50 ★

nearly 40-year life of the program, which is now part of the Status and Trends of Biological Resources Program at the USGS.

Information collected through the BBS led to research and management actions to help neotropical migratory songbirds in the late 1980s and grassland species in the 1990s. Since 1991, grassland birds have had fewer species with increasing population trends than any other bird group examined in North America. Unfortunately, the decline of most grassland bird populations continues today, as shown by declines of 77 percent for grasshopper sparrows and 66 percent for eastern meadowlarks between 1966 and 2003.

Ground Water—The Invisible Resource

Today, 130 million people drink ground water, but much

less is known about the quantity and quality of our ground water than about surface water. In 1900, there were only three ground-water scientists in the USGS. Throughout the 20th century, that number increased as awareness of the importance of this essential component of our Nation's water endowment grew. Studies in the early 1900s on regional ground-water systems throughout the Nation, such as those undertaken in the central Great Plains, the Los Angeles region, and Long Island, laid the foundation for O.E. Meinzer's first assessment of the ground-water resources of the entire United States in 1923.

State-by-State summaries in midcentury provided more information, and a series of publications beginning in the late 1970s evaluated 25 of the Nation's most important regional ground-water systems by using a quantitative evalu-

California Department of Water Resources

ation that led to many innovations in modeling these systems. USGS contributions to developing quantitative models have been critical to understanding such technical topics as well hydraulics—how water flows to wells—and systems analysis—studying the sources of water to an aquifer on a systemwide basis rather than at single wells. The use of computer models has greatly enhanced the ability of scientists to understand and predict movement of water underground, and the MOD-FLOW model developed by USGS scientists in the 1980s

Oscar E. Meinzer

In 1906, when Oscar E. Meinzer was first employed by the USGS, the science of ground water was a little-recognized field. His persistence and diligence over the 41 years of his career would change that. Meinzer's sincere dedication had so great an influence on the scientific community that he has become known as the father of ground-water hydrology, and the period of 1910 to 1940 has been labeled the Meinzer Era.

Meinzer began his career as a Junior Geologist and advanced quickly to be the third USGS Ground-Water Division Chief. Under his leadership, a systematic scientific approach was applied to the problems of hydrogeology, and the underlying principles were defined. His personal scientific efforts, insight, and leadership greatly influenced the direction and development of the science of hydrogeology, both inside and outside the USGS; by the time of his death in 1948, Meinzer had achieved international recognition as being preeminent in ground-water science.

Advancing the field of hydrogeology was not always an easy path. Meinzer once said that much of his work had been done almost surreptitiously because of a lack of funds and government interest. He worked around many roadblocks with firm principles and a strong belief that hydrology was a vital public service. In his introduction to the 1942 volume "Hydrology," which he edited, Meinzer sums up his viewpoint saying, "The science of hydrology is thus intimately connected with the development of human society. In each project, advance in hydrology has come in response to the needs of the people, and each advance in the science has made possible more effective service."

1930–1943

Walter Curran Mendenhall is the 5th Director of the USGS. President Hoover appoints Mendenhall to succeed Smith as Director of the USGS, honoring his commitment to appoint the heads of scientific agencies from within the civil service. Mendenhall had made notable contributions to the geology of Alaska, and his study of ground-water hydrology established it as a field of scientific endeavor.

1939

The Bureau of Fisheries (from the Department of Commerce), and the Bureau of Biological Survey are moved to DOI.

1940

DOI combines the Bureau of Fisheries and the Bureau of Biological Survey to form the U.S. Fish and Wildlife Service (USFWS).

The USGS Library

The U.S. Geological Survey Library is the largest library for earth sciences in the world. Since its establishment in 1879, it

has acquired more than one million books and journals, one million maps, 370,000 microforms, 270,000 pamphlets, and 250,000 photographs. It has been a catalyst for cooperation, understanding, and new discoveries, as scientists around the world share and study the results of past scientific investigations, current trends and techniques, and new research directions.

The USGS library system is managed by the Geospatial Information Office (GIO) and includes libraries in Reston, Va.; Denver, Colo.; Menlo Park, Calif.; and Flagstaff, Ariz. There are also libraries associated with science centers and field offices across the country holding specialized collections related to the research of their local scientists. Information describing library activities can be found at *<library.usgs.gov>*.

The Library collects, classifies, catalogs, and preserves natural science publications in many formats and languages as resources for present and future use by students, the general public, teachers, technical and scientific professionals, USGS staff, and policy- and decisionmakers at all levels of government. Libraries throughout the world, including the largest and most renowned, borrow from the USGS Library's unique collection, and users of the USGS Library have the reciprocal benefit of being able to borrow materials from other libraries.

Despite its age, the USGS Library is keeping up with the times by heading toward a "virtual library" where users will have desktop access to an increasing wealth of resources 24 hours a day, 7 days a week.

is considered by many to be the most used tool worldwide for quantifying ground-water flow systems. MODFLOW has continued to evolve as new understanding is gained and as our Nation increasingly relies on ground water for drinking water and other uses. Today, roughly 500 USGS ground-water hydrologists across the Nation are studying this essential resource. A growing awareness of the interdependence of surface water and ground water is leading to new insights that will help ensure the Nation has the water it needs for a growing economy and a growing population.

Understanding and Classifying Wetlands

Historically, wetlands have been regarded as swampy lands that bred disease, impeded agriculture and development, and restricted overland travel. The wetland policy of the U.S. Government beginning in 1849 was elimination. It has been estimated that more than half of the naturally occurring wetlands in the lower 48 States have been lost since the 1700s. Not until the last quarter of a century did society begin to understand the value of wetlands for wildlife habitat, flood attenuation, water-quality improvements,

and sediment removal. As this understanding has grown and wetlands policy has changed, resource managers are finding that classifications or definitions of different wetland types are useful for making decisions with respect to wetland regulation and protection, understanding restoration processes in wetland types, and identifying the types of wetlands that are more valuable or more threatened in a given region.

Early wetland classification systems in the United States were motivated by interest in converting wetlands to croplands and by the need to differentiate wetlands from other land-cover types for regional and national planning purposes. Scientists from the Fish and Wildlife Service, led by Lewis Cowardin, worked with wetland scientists and mapping experts from the USGS, the National Oceanic and Atmospheric Administra-

1943–1956
William Embry Wrather is the 6th Director of the USGS.

1946
The Bureau of Land Management is created out of the General Land Office and the USGS.

1949–1954
The Aleutian Volcano Observatory is established.

1950
Public Inquiries Offices (PIO) are established in Denver and Salt Lake City. Seven more PIOs open over the next few decades. By 2004, these offices are being transformed into an integrated Natural Science Network to make USGS data, information, and knowledge available to anyone, anywhere, at any time.

tion (NOAA), and academia to develop a new classification system based on the ecosystem concept. A draft of the "Classification of Wetlands and Deep-water Habitats of the United States" by Cowardin and others was extensively field tested and reviewed before it was published in 1979. The objective of this classification system is to impose boundaries on natural aquatic ecosystems for inventory, evaluation, and management purposes. The major systems (marine, estuarine, and others) are distinguished by a variety of hydrologic, geomorphologic, chemical, and biological characteristics. Subsystems, classes, and other categories are defined by vegetation, water chemistry, soil, and other characteristics.

The Cowardin classification system became the basis for the FWS National Wetland Inventory's Wetland Status and Trends report and is now used by Federal and State Governments as the basis for wetland classification. Wetlands research has increased our understanding of how wetlands function, where they are found within the landscape, and how the interactions among hydrology, soils, and vegetation differ in different wetland types. USGS scientists are investigating wetland processes to better understand the steps involved in sustainable restoration and to make better decisions in managing wetlands for migratory waterfowl use, flood attenuation, hydrologic connections, and the role of wetlands in carbon sequestration and sea-level rise.

Mapping Beneath the Sea

The U.S. EEZ (Exclusive Economic Zone) was declared by Presidential Order in 1983. This area includes the sea floor extending 200 nautical miles away from all U.S. possessions and trust territories. This act added more than 3 million square nautical miles to the United States, an area larger than the 3.6 million square miles of U.S. onshore lands. At this time, the general bathymetry of the EEZ was known, but the detailed physiography was not. The USGS, charged with surveying this new domain, launched a program in 1984 using a long-range sidescan sonar system (GLORIA) to study the entire EEZ. The acoustic images produced by the program are as remarkable as the first photographs from the far side of the Moon. The images show a wealth of geologic features, including volcanic structures, channels, and scars from submarine landslides. GLORIA imagery of the U.S. EEZ was the first systematic mapping of sea-floor

Menlo Park Center

In January 1954, 120 employees moved into what would be the first of many USGS buildings in Menlo Park, Calif. These USGS employees were brought together in a centralized Western Region facility to increase scientific cooperation and efficient use of resources. From a modest beginning of geologic-mapping and mineral-resource activities in 1954, the Menlo Park Center expanded rapidly until it included 2,000 people in almost two dozen buildings. The Center added major research efforts in earthquake and volcano hazards, geothermal energy, hydrology, isotope geology, marine geology, coastal processes, and ecosystem studies, as  well as the largest earth science library west of Reston, Va., and a major marine facility at the Port of Redwood City. Recent decentralization has reduced the USGS presence in Menlo Park to approximately 600 people. Today, the USGS facility in Menlo Park is a world-renowned center of scientific excellence and the largest and most diverse earth science facility west of the Rocky Mountains. As the USGS celebrates 125 years as a Federal science agency, it is also celebrating 50 years of scientific achievements in Menlo Park. Highlights include 35 years of monitoring water quality in San Francisco Bay and developing expertise in estuarine ecosystems; establishing an earthquake research program in 1965; leading the production of digital orthophoto imagery for the Nation; critical understanding of heat flow and permafrost leading to fundamental design changes in the 1970s for the planned Trans-Alaska petroleum pipeline; and, in the 1960s, developing the paleomagnetic time scale, a fundamental piece of evidence in the theory of plate tectonics. In the 1980s, Menlo Park developed creative new programs in education and public outreach, establishing regular open houses and becoming a model for the rest of the bureau. For more of the history of USGS at Menlo Park, please see *<menlocampus. wr.usgs.gov/50years>.*

1954

The USGS is 75 years old. The USGS has 7,000 employees and appropriated funds of $27 million, with total funds, including those from States and other agencies, of nearly $48 million. Streamflow data are obtained from 6,400 gaging stations, and the chemical quality of more than 85,000 samples of water is examined. New technology and aerial photographs and photogrammetric methods result in significant increases in the amount of mapping accomplished. Geologists adapt photogrammetric methods to their mapping and apply modern statistical methods to field geology. The USGS has the responsibility for supervising more than 100,000 lessee operations on mining or oil-and-gas properties on Federal lands and the Outer Continental Shelf.

75

William Thomas Pecora

William T. Pecora, Director of the USGS from 1965 to 1971, is largely responsible for two major contributions to public health, public safety, and public prosperity: the creation of a national earthquake research center at the USGS facility in Menlo Park, Calif., and the Landsat series of Earth-observing satellites.

Pecora joined the USGS in 1939. For several years, he investigated strategic-mineral deposits in the United States and Latin America and then engaged in a long-range study of rare mineral deposits in Montana.

Named Chief Geologist in 1964, Pecora collaborated with NASA, enabling the USGS to accelerate its remote-sensing research in analyzing the potential values of surveying the Earth from space. Pecora carried a broad vision and deep appreciation of the use of satellite system programs for continuous inventory and management of our national resources. He was a motivating force behind the establishment of Earth remote sensing from space, and it was under his leadership that the Landsat series of Earth-observing satellites became a reality.

In response to the great Alaska earthquake of 1964, he obtained approval to establish the National Earthquake Research Center in Menlo Park. The Center provided a focus for research on the causes and effects of earthquakes and on methods to predict the time and location of destructive earthquakes.

In 1965 Pecora was elected to the National Academy of Sciences and appointed Director of the USGS. In April 1971, he left the USGS to become Under Secretary of the Interior and served until his death in 1972, just a few days before the launch of the first Landsat satellite.

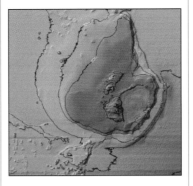

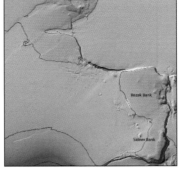

morphology and features in the deep ocean, and its value has been recognized by the entire marine community. Results from the GLORIA program opened the door to exploration for the mineral and energy resources found on and under the sea floor.

More recent work has focused on habitat and sea-floor maps of National Marine Sanctuaries to serve as a basis for managing sanctuary resources and for conducting research. Mapping of seabed habitats is a developing field that has required the integration of geologic and biologic studies and the use of imaging techniques

such as multibeam and sidescan sonar. To compile interpretive maps showing seabed environments and habitats, USGS scientists developed a sea-floor classification system that is the basis for comparing, managing, and researching characteristic areas of the seabed. Seabed maps of the sanctuaries are being used for management and research decisions that affect commercial and recreational fishing, habitat disturbance, engineering projects, tourism, and cultural resources.

The USGS has embraced airborne laser mapping (LIDAR) for probing the sea-floor; this tool enables cost-

effective, precise, and rapid mapping of coastal topography and nearshore bathymetry. USGS coastal scientists are using this technique to assess post-storm coastal damage and change from hurricanes. The USGS responded to the impact of Hurricane Isabel in North Carolina in September 2003 through a series of meetings with managers from the Federal Emergency Management Agency (FEMA) and North Carolina State emergency agencies. USGS data allowed the agencies to design appropriate responses to the emergency; for example, the Department of Transportation used the data to help design and rebuild the road infrastructure that was lost during Hurricane Isabel. As population concentrates along the coastlines, understanding and characterizing issues of coastal erosion and wetland loss will become increasingly important.

1956–1965

Thomas Brennan Nolan is the 7th Director of the USGS. Nolan leads the USGS to a broadened and intensified commitment to basic research, to the advancement of geology in the public service, and to the prompt publication of USGS results.

1958

David Varnes develops a vocabulary of terms to describe landslides, thus giving landslide experts a common language. This system of naming landslides first by their material then by the type of movement is adopted throughout the world.

Geographic Tools for the Nation

As the Nation grew and changed, the earlier emphasis on mapping for resource development broadened to include resource protection and conservation. Through the mid-20th century, the USGS focused on providing detailed coverage of the Nation's topography through a series of maps at 1:24,000 scale, but there were also increasing demands for other kinds of geographic and cartographic information. Changes in public policy, customer usage, and technology spurred the development of new products of broad geographic information and reference such as digital terrain maps, aerial photography, and satellite imagery. The acclaimed "National Atlas of the United States," published in 1970, was the Nation's first official atlas. It included more than 700 printed maps

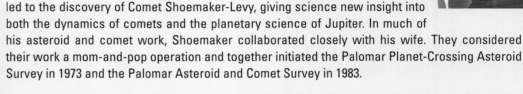

Eugene M. Shoemaker

Born in Los Angeles, Calif., in 1928, Gene Shoemaker was a visionary geologist. In his early career, he believed geologic studies would be extended into space and dreamed of being the first geologist to map the Moon. Though he never actually got to the Moon, he created the Branch of Astrogeology within the USGS, established the USGS Field Center in Flagstaff, Ariz., and led research teams investigating the Moon.

Shoemaker, known as the "Father of Planetary Geology," influenced the geosciences in numerous ways. He provided the definitive work on basic impact cratering and discovered a number of new craters on Earth. With the help of his wife Carolyn (a planetary astronomer) and David Levy, a decade-long survey for Earth-crossing asteroids and comets

led to the discovery of Comet Shoemaker-Levy, giving science new insight into both the dynamics of comets and the planetary science of Jupiter. In much of his asteroid and comet work, Shoemaker collaborated closely with his wife. They considered their work a mom-and-pop operation and together initiated the Palomar Planet-Crossing Asteroid Survey in 1973 and the Palomar Asteroid and Comet Survey in 1983.

He once said he considered himself a scientific historian whose mission in life was to relate geologic and planetary events in a perspective manner—a modest statement from a man who almost single-handedly created planetary science as a discipline distinct from astronomy. His work resulted in more than three decades of discoveries about the planets and asteroids of the solar system. He helped train the Apollo astronauts and sat beside Walter Cronkite providing commentary during the moon walks. His moon studies culminated in 1994 with Project Clementine, for which he was the science team leader. Among numerous awards, he received the National Medal of Science, the highest scientific honor bestowed by the President of the United States.

1961
Eugene Shoemaker founds the Astrogeology Research Program and becomes its first Chief Scientist. He establishes the Flagstaff Field Center in 1963.

1962
The USGS begins a program of marine studies to identify and evaluate potential mineral resources on or beneath the sea floor and to aid in solving the problems caused by rapid population growth, urbanization, and industrial expansion in coastal areas.

1963
The USGS, in cooperation with NASA, begins to train astronauts in geology and to investigate and evaluate methods and equipment for exploration of the Moon.

1963
The USGS Mission in Saudi Arabia is formally established, 19 years after the first USGS geologist arrived to advise on the natural resources of the Kingdom.

1964
For the first time, the appropriations and total funds available for the fiscal year beginning July 1 exceed $100 million. This is more than double the amount of just a decade earlier.

James R. Balsley

James R. Balsley, Jr., was one of those rare individuals who is both a first-rate scientist and an exceptional manager. He is fondly remembered for his abiding concern for the welfare of the people doing the science, as well as for the science itself. During his career, he contributed much to the field of geology and the long-term scientific health of the USGS.

Balsley started at the USGS in 1939, mapping geology in Washington State's Olympic Peninsula. During World War II, he began to specialize in geomagnetism as he worked on geophysical investigations for Federal civilian and military organizations. This work led to aeromagnetic surveys of Antarctica and Alaska in the late 1940s. He developed new geophysical techniques to improve mapping of deep geologic structures and to detect hidden ore bodies. In the 1950s, as Chief of the Branch of Geophysics, Balsley established the Rock Magnetism Project, a project that led to key discoveries supporting the theory of plate tectonics.

Balsley was a strong proponent for applying USGS science to societal problems. He established the Land Information and Analysis Office, where staff worked directly with land-planning organizations, Federal and State hazard coordinators, and other groups to translate and apply USGS science to their needs. This activity led to a new role in environmental studies for the USGS that continues today. As Assistant Director, Balsley successfully coordinated USGS research programs with those of most other agencies in the Federal Government.

Balsley left a legacy of outstanding scientific imagination and exceptional managerial creativity. His strong emphasis on applying the results of USGS science to societal issues remains a fundamental operating principle of the USGS today.

showing the physical, historical, economic, and sociocultural landscape of the Nation.

A national program for consistent, multiuse aerial photography was established in 1987. The National Aerial Photography Program is an interagency Federal effort coordinated by the USGS, covering the lower 48 States and Hawaii. Photographs are acquired from airplanes flying at an altitude of 20,000 feet. The program was originally developed to assist in updating USGS topographic maps, but other Federal and State agencies have found these photos useful for their needs as well. USGS has provided leadership for over 30 years in the Landsat program of land remote sensing. The USGS archive of remote-sensing data at the EROS Data Center in Sioux Falls, S. Dak., houses millions of images of the Earth and represents a priceless resource for science and research for years to come.

The Geographic Names Information System (GNIS), developed by the USGS in cooperation with the U.S. Board on Geographic Names, contains information for almost 2 million physical and cultural geographic features in the United States. The Federally recognized name of each feature described in the database is identified, and references are made to a feature's location by State, county, and geographic coordinates. The GNIS is our Nation's official repository of domestic geographic names information.

1964

The Prince William Sound, Alaska, earthquake of magnitude 9.2 results in massive destruction in Anchorage. USGS geologists are assigned to the Task Force of the Federal Reconstruction and Development Planning Commission to select sites for rebuilding. The establishment of the USGS's Center for Earthquake Research in Menlo Park, Calif., follows.

U.S. Army

1964

Congress passes the Wilderness Act, setting more than 9 million acres of national forest lands aside from permanent roads, buildings, and commercial activities. The USGS and the U.S. Bureau of Mines are asked to assess the mineral resources in each area of the proposed or established wilderness by the end of 1983.

1965

Scientists at the Patuxent Wildlife Research Center play a role in crafting the Endangered Species Preservation Act, and research on endangered species begins as an effort to recover the whooping crane from near extinction.

1965–1971

William Thomas Pecora is the 8th Director of the USGS. Pecora leaves the USGS in 1971 to become the Under Secretary of the Interior.

Challenges and Opportunities Today

National Assessments— Looking at the Land

Natural processes—the flow of rivers, the migration of birds, the ground shaking triggered by an earthquake—are not constrained by political boundaries. The USGS has the unique ability and duty to conduct large-scale regional and national assessments that provide an accurate portrait of different aspects of our land. For example, in the early 1970s, visionary USGS scientists like William Pecora and James Anderson established the capabilities to routinely use instruments on orbiting satellites to map the Nation's land use and land cover, the biophysical pattern of natural vegetation and agriculture, and wilderness and urban areas. The USGS leads a nationwide effort to map land cover, working through a consortium of Federal partners. New technologies applied to remote-sensing data, including computer modeling and GPS, give land-cover analysts powerful tools to deliver consistent, current nationwide data that address a wide variety of issues, from planning and development to natural hazards and the environment.

The USGS conducts nationwide water-quality assessments designed to answer, in a scientifically defensible way, the question, "What is the quality of the Nation's water?" The National Water-Quality Assessment Program assesses the quality of streams and ground water in more than 50 major river basins and aquifer systems across the Nation—the source for more than 60 percent of the Nation's drinking water and water for irrigation and industry. Each assessment follows a nationally consistent study design and methodology, thereby providing information about local water-quality conditions that can be compared regionally and nationally to provide insight on where and when water quality varies and on the influence of human and natural factors on water quality. Water-quality professionals at local, State, Tribal, and national levels use this information to develop strategies for managing, protecting, and monitoring water resources in many different hydrologic and land-use settings across the Nation. This information can help identify key causes of nonpoint-source pollution in agricultural and urban areas, contribute to State assessments and planning, sustain the health of aquatic ecosystems through improved stream protection and restoration management, and other tasks.

Determining the status (abundance, distribution, productivity, and health) and trends (how they are changing over time) of our Nation's living natural resources is critical to facilitate research, enable resource management and stewardship, and promote public understanding and appreciation of our living resources. The USGS Status and Trends of Biological Resources Program responds to these needs through inventory and monitoring efforts at multiple biological, spatial, and temporal scales. For example, to support recovery efforts of threatened and endangered species such as the grizzly bear, hair samples are collected and analyzed for their genetic makeup and then used to estimate the size of the

---

1966
The North American Breeding Bird Survey, a program coordinated jointly between the USGS and the Canadian Wildlife Service, begins as a long-term, large-scale, international avian monitoring program to track the status and trends of North American bird populations.

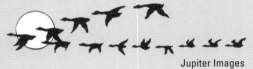

Jupiter Images

1966
DOI establishes the Earth Resources Observation Systems (EROS) Program in Sioux Falls, S. Dak.

1970
The National Atlas of the United States of America, the Nation's first official atlas, is published. It describes the Nation extensively, with more than 700 physical, historical, economic, sociocultural, and administrative maps.

1970s
The National Wetlands Research Center is the first to document the unique and tremendous wetland loss in Louisiana.

Jayne Belnap

Jayne Belnap leads the sort of interesting, accomplished, and well-traveled life many of us dream about. She studies how different land uses—including some of our favorite recreational activities, such as hiking and biking, and activities that support our society, such as livestock grazing and energy exploration—affect the fertility and stability of desert soils.

Applying that knowledge, she then studies the factors that make some desert communities susceptible to invasion by exotic plants, while others remain uninvaded, despite similar use patterns. Using repeat photography, she has also examined how plant communities in the West have changed over the past 100 years.

Her studies have taken her to a number of exotic places, including South Africa, Kenya, Zimbabwe, Mongolia, China, Siberia, Australia, and Iceland, where she advises scientists and managers on how to maintain soil fertility and stability while still using the land. She also travels extensively within the United States, interacting with and training Federal, State, and private land managers on how to best manage dryland ecosystems.

Belnap has been a scientist with the Department of the Interior since 1987 and is currently with the USGS in Moab, Utah. She is the past Chair for the Soil Ecology section of the Ecological Society of America, is President-Elect of the international Soil Ecology Society, is a subject editor for the journal Ecological Applications, and participates in a number of other professional functions.

grizzly bear population. To satisfy international agreements and treaties, guide the establishment of fishing limits, inform management, and facilitate our understanding of fish communities and their interaction within the Great Lakes, annual fish trawl surveys in all five Great Lakes are conducted. The Status and Trends program periodically assembles the information from these and other efforts to then publish comprehensive reports that synthesize our understanding of the Nation's biological and ecological resources.

Ecosystem Assessment and Restoration

Working with other Federal partners, States, and local governments, the USGS is helping to provide data and scientific understanding to assess the water and habitat quality of crit-

ical ecosystems and to analyze the effectiveness of restoration activities. The Chesapeake Bay, the San Francisco Bay, and the Platte River Valley are among the ecosystems under study.

The Chesapeake Bay, the Nation's largest estuary, has historically supported one of the most productive fisheries in the world, and the bay's watershed provides vital habitat for migratory birds using the Atlantic

Flyway. Unfortunately, the commercial, economic, and recreational value of the bay and its watershed has been degraded by poor water quality, loss of habitat, and overharvesting of living resources. The restoration of the bay and its watershed is overseen by the Chesapeake Bay Program, which is a partnership among States, the District of Columbia, Federal agencies, and the Chesapeake

1970s
The Cooperative Park Study Units, a university-based program of scientific studies on the ecological, environmental, and sociological aspects of park and wildland management, begin, benefiting both the parks and the universities.

1970s
USGS scientist Russell Campbell develops the concept for documenting rainfall duration and intensity as it is associated with shallow landslides. The concept will later be refined by Susan Cannon, Stephenson Ellen, Gerald Wieczorek, and others and will be essential for the near-real-time landslide warning systems that the USGS will operate in cooperation with State and local governments.

Bay Commission. Since the late 1970s, the USGS has provided unbiased scientific information that is used to help formulate, implement, and assess the effectiveness of restoration goals to reduce nutrient and sediment to the bay and improve habitats for fish and migratory birds in the bay and its watershed.

Since 1849, the San Francisco Bay area has been altered to meet demands for agricultural and urban land, freshwater, and the riches supplied by gold deposits in the watershed. Tidal marshes have been filled, freshwater has been diverted, and many nonnative species have been imported and released, whether intentionally or by accident. Today, there are few commercial fisheries in the system, several of the remaining native fish are endangered or threatened, the contamination of recreationally important sport fishes is sufficient to limit consumption by people, and the need for more freshwater is a critical component of the political landscape of California and other Western States. The USGS is a founding member of the Interagency Ecological Program for the San Francisco Bay and freshwater delta, a group of 10 members including State, Federal, and nongovernment organizations. USGS studies have focused on water quality and on how tidal waters and freshwater flow through the system. New projects in San Francisco Bay will evaluate the success of marsh restoration and study how climate is linked to watershed processes and freshwater flow, providing numerical models of the system to State and Federal water managers.

The unique land and water resources of the central Platte River Valley created a natural "highway" for people and wildlife in Nebraska, along which half a million sandhill cranes and several million other waterfowl migrate each year. The valley also provides living space for nine endangered species, including the whooping crane, piping plover, and interior least tern. Changes in water management, sediment levels in the river, and land use have changed the river channel and nearby wet meadows, reducing their ability to support migratory and resident birds and other plants and animals. USGS scientists are continuously monitoring Platte River streamflow and

National Wildlife Health Center

Occasional reports of large-scale wildlife disease events surfaced in the early 1900s. By the early 1970s, there were still only a few laboratories where wildlife disease was the subject of serious scientific inquiry. In January 1973, however, a die-off of 40,000 mallards at the Lake Andes National Wildlife Refuge in South Dakota brought wildlife disease into the national spotlight. Two years later, the National Wildlife Health Center (NWHC) was established in Madison, Wis., launching a nationwide wildlife health program. The laboratory's first director, Milton Friend, is internationally recognized for  his accomplishments in the fields of wildlife health and wildlife disease ecology; his leadership of the wildlife health program set a standard for excellence that has been consistently maintained. Wildlife health touches a broad range of issues including public health and domestic animal health; the focus on wildlife as a function of ecosystem health is a specialty of the NWHC. Understanding the role of disease agents in ecosystems is critical to managing natural resources and preserving ecosystem function under ever-increasing demands on our planet's natural resources. NWHC research on zoonotic diseases (those that can be transferred between animals and humans) concentrates on improving our understanding of the ecological relationships among free-ranging wildlife, domestic animals, and humans. The Center works closely with other Federal and State agencies and Native communities to provide dedicated field support and exceptional diagnostic and research capabilities that can be applied to manage and prevent disease outbreaks.

Through its commitment to scientific excellence, the NWHC has become the most comprehensive facility in the world dealing specifically with wildlife health and disease. It has achieved a national and international reputation, emerging as a leader in wildlife health through preeminent science, innovative technology, and responsive service.

1971–1978

Vincent Ellis McKelvey is the 9th Director. McKelvey, a former Chief Geologist and career USGS scientist, heads an organization that has an operating budget of $173 million and 9,200 employees. General-purpose topographic maps are available for 84 percent of the total area of the 50 States and territories, streamflow data are collected at more than 11,000 gaging stations, and mineral production from lands supervised by the USGS is valued at more than $3 billion. McKelvey's term is marked by an increase in multidisciplinary studies and in the complexity of USGS operations.

Timothy L. King

To most of us, a salmon is a salmon, but to Tim King, a Research Fishery Biologist with the USGS, there are considerable and important differences between the salmon of different areas. King conducts molecular genetics research on threatened and endangered species, and his job is to understand diversity in the smallest of groupings. He focuses on the development of DNA markers that allow assessment of population genetic structure, phylogeography, and phylogenetics. His goal is to provide natural resource managers with the guidance they need to identify appropriate units of conservation. King headed a group of researchers who established the existence of distinct genetic differences between North American and European Atlantic salmon and between Maine and Canadian salmon. Considerable differentiation was also identified among salmon inhabiting streams of coastal Maine. King's Atlantic salmon research was reviewed and unequivocally supported by a Congressionally mandated National Academy of Sciences panel of geneticists. In case you think there's something fishy about his research area, King has also made significant contributions to the understanding of the biology and population structure of Atlantic sturgeon, brook trout, spotted salamanders, wood frogs, bog turtles, horseshoe crabs, and black bears.

water temperature and transmitting the data to publicly accessible World Wide Web pages; using satellite tracking to follow individual cranes throughout their intercontinental migration routes; and partnering with the National Aeronautics and Space Administration (NASA) to develop remote-sensing techniques that provide quick and accurate data to researchers assessing changes in channel shape and determining the effects of riverbank changes on ecological communities.

America's Amphibians: Vanishing from the Landscape?

Amphibian populations throughout the world, including frogs and salamanders in the United States, have declined markedly over the past 20 years, but the reasons for these declines, as well as for a growing number of malformed individuals, are poorly understood. In 2000, the USGS began the Amphibian Research and Monitoring Initiative (ARMI) to monitor trends in amphibian populations and learn more about the causes of their declines. Amphibians have many unique or unusual development characteristics. For example, most amphibians spend part of their lives in aquatic habitats; others can breathe through their skin. These characteristics make them more susceptible to environmental changes, such as the loss of aquatic habitat and the presence of environmental toxins that are easily absorbed through the skin.

Understanding how these unique characteristics relate to population declines requires a multidisciplinary approach. Biologists study life cycles, habitat interactions, and population trends. Hydrologists and geographers investigate surface water and the landforms that influence the size, flow, and distribution of water bodies. Cartographers bring an understanding of spatial relationships. Chemists and toxicologists study the effects of contaminants on individual animals and populations. Many possible causes of amphibian decline have been proposed recently. Declines in some amphibian populations may be the

1972

The U.S. Environmental Protection Agency bans DDT after Patuxent Wildlife Research Center scientists discover that DDT, in addition to causing bird deaths, can cause other reproductive failures.

1972

A cooperative program involving the USGS, NASA, and universities to systematically map the geology of Mars results in a shaded-relief map of Mars.

1973–1974

In 1973, the USGS moves its National Headquarters to a new building in Reston, Va. The building, the John Wesley Powell Building, is formally dedicated on March 3, 1974, as the fifth National Center. The USGS takes on the responsibility for operational research in seismology and geomagnetism by agreement with NOAA, and part of NOAA is transferred to the USGS.

result of natural fluctuations, but some of the declines are likely the result of the introduction of a contaminant or alteration of a habitat. Regional phenomena like changes in climate, introduction of nonnative competitive species, or exposure to pathogens may also play a part. Many herpetologists believe that a combination of stresses is being placed on amphibian populations and that the combination of stresses puts some species at risk more than others. Rigorous long-term studies of amphibians are needed to assess the magnitude and direction of changes in amphibian populations. Interagency partnership is a critical component of the ARMI program. The USGS is partnering with Federal and State agencies, herpetological societies, conservation advocacy organizations, and educational institutions to find the causes and the solutions for our disappearing amphibians.

Tracking Down Biological Information

Information on local, regional, State, and global biology is available from thousands of databases across the landscape. Making this wealth of existing information accessible is the challenge that led to the creation of the NBII—the National Biological Information Infrastructure. NBII is a broad, collaborative program to provide increased access to data and information on the Nation's biological resources. Resource managers, scientists, educators, and the general public use the NBII to answer a wide range of questions related to the management, use, or conservation of this Nation's biological resources. It is an information pipeline coordinated by the USGS, whose contents are supplied by partners in Federal, State, and local government agencies, nongovernmental

organizations, academia, private industry, and international institutions. The NBII links diverse, high-quality biological databases, information products, and analytical tools maintained by NBII partners and other contributors. NBII partners and collaborators also work on new standards, tools, and technologies that make it easier to find, integrate, and apply biological resources information.

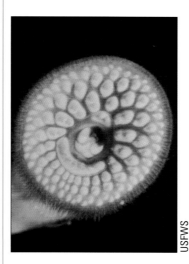

USFWS

L. David Mech

Jerry Sanders

Dave Mech wasn't raised by wolves, but after reviewing his career, you might think so. He has studied wolves and their prey in northeastern Minnesota since 1968; in Yellowstone since 1995; and in Denali National Park, Alaska, from 1986 to 1995—not to mention that each summer from 1986 to 1996 he lived with a pack of wolves in Canada's High Arctic, studying their behavioral interactions and their predation on musk oxen and arctic hares. He continues to study wolves and their prey there each summer.

Mech is a Senior Research Scientist with the Northern Prairie Wildlife Research Center and an Adjunct Professor at the University of Minnesota in St. Paul. He began working for the Department of the Interior as a Wildlife Research Biologist for the Fish and Wildlife Service in 1968.

He has chaired the Wolf Specialist Group of the World Conservation Union (IUCN) since 1978 and is the founder and vice chair of the International Wolf Center in Ely, Minn. Mech has been involved with many international projects, including work on wolves in Italy, Spain, Portugal, and Croatia; leopards and lions in Kenya and Tanzania; and tigers, elephants, lions, and other species in India. His many awards include The Wildlife Society's Aldo Leopold Award for Distinguished Service to Wildlife Conservation.

For those of us who would like to know more about wolves and other wildlife but aren't quite ready to spend our summers living with them, Mech has published 10 books and more than 300 scientific and popular articles. His latest book, a comprehensive reference work, titled "Wolves: Behavior, Ecology, and Conservation," was published in September 2003 and is now in its second printing.

1976
The USGS Library closes the traditional card catalog and establishes a digital catalog.

1977
Congress passes the Earthquake Hazards Reduction Act of 1977, to reduce the loss of life and property from future earthquakes.

1977
Congress directs the USGS to establish a national water-use information program. It becomes part of the Federal-State cooperative program.

1978
The National Seismic Network is established.

1978
The Conterminous U.S. Minerals Assessment Program is created.

1978
USGS joins the Advanced Research Projects Agency Network (ARPANet), a forerunner of the internet.

Cooperative Research Units

Every year, USGS scientists advise and mentor more than 670 graduate students as part of USGS participation in the Cooperative Research Units program. The program was established in 1935 to train personnel in the rapidly growing field of wildlife management, and later fisheries, and to provide better technical information for professional wildlife and fisheries managers. It has become one of the USGS's strongest links to Federal and State land and natural-resource management agencies.

Bill Pine

The program is a unique model of cooperative partnerships among Federal and State Governments, academia, and the Wildlife Management Institute. Today, there are 40 Cooperative Fish and Wildlife Research Units located on university campuses in 38 States. By pooling resources, everyone benefits: the USGS provides Federal research scientists; the universities provide office space, administrative support, and access to university facilities; the State game and fish agencies provide base funding and logistical support for research activities; and the Wildlife Management Institute helps to coordinate the program at a national level.

Unit scientists and their students conduct applied research on current fish and wildlife issues to address the needs of State and Federal cooperators and partners. Their projects range in scale from microscopic to landscape and cover a wide range of topics on game and nongame animals and their habitats. Unit studies address State and Federal agency information needs on biodiversity and conservation biology, migratory birds, inland fisheries, anadromous fish, habitat assessment, population estimation and modeling, and numerous other topics. Scientists and students provide technical assistance to State and Federal agencies, helping them to apply the research findings to their specific needs in areas as diverse as Everglades National Park in Florida and the Arctic National Wildlife Refuge in Alaska.

USDA, Forest Service

The NBII is organized in nodes. The California Information Node, for example, provides data discovery, exchange, and interoperability for resources critical to the study of California's diverse ecosystems, with an initial emphasis on invasive species issues. The Southern Appalachian Information Node focuses on providing access to data and information that will facilitate tracking, understanding, and predicting changes in biodiversity in Tennessee, Kentucky, Alabama, and Mississippi, with a particular interest in measuring and mitigating human impacts through technology and education. The Fisheries and Aquatic Resources Node provides an extensive array of data on managed fish strains, populations, and broodstocks located throughout the United States, as well as watershed-based data and extensive State-by-State fishing resources.

Bureau of Land Management

The Gap Analysis Program (GAP) was created to provide clear, geographically explicit information on the distribution of native vertebrates, their habitat preferences, and their management status, information needed to complement species-by-species management in the context of large-scale loss of habitat. The program goal is to develop predictive information that can be used to manage the Nation's biodiversity so that ordinary plant and animal species will not become threatened with extinction. GAP promotes the conservation of biodiversity by developing and sharing information on where species and natural communities occur and how they are being managed for long-term survival. It is an important part of the NBII. GAP began by mapping distributions of amphibian, bird, mammal, and reptile species. The program is currently developing methods to extend coverage to fish, mussels, crayfish, snails, and other species and will expand to more species as knowledge and resources allow. GAP projects rely on the participation of nearly 500 cooperating State and Federal agencies, academic and nonprofit institutions, and private industry.

Monitoring at Many Levels

Remotely sensed data are fundamental tools for studying the Earth's land surface, including coastal and nearshore environments. The USGS has been a leader for decades in providing remotely sensed data to the national and international communities. Stemming from its historical topographic mapping mission, the USGS has archived and distributed aerial photography of the United States for more than half a cen-

1978–1981
H. William Menard is the 10th Director of the USGS.

1979

Menard's first year in office includes the centennial of the USGS. The appropriation is about $640 million, and total funds available are $765 million. The USGS staff numbers more than 12,000 (including contractors). The USGS is one of only a few Federal agencies to survive for 100 years with its original name and mission unchanged.

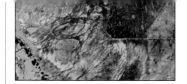

tury. Since 1972, the USGS has acquired, processed, archived, and distributed Landsat and other satellite and airborne remotely sensed data products to users worldwide. Today, the USGS operates and manages the Landsat 5 and 7 satellite missions and cooperates with NASA to define and implement future satellite missions that will continue and expand the collection of moderate resolution remotely sensed data.

Remotely sensed data are critical sources of information for studying sources for energy and minerals, coastal environmental surveys, assessments of natural hazards (earthquakes, volcanoes, and landslides), biological surveys and investigations, water resources status and trends analyses and studies, and geographic and cartographic applications such as wildfire detection and tracking. The archive of remotely sensed data is used by the global sci-

ence community to evaluate manmade and natural changes to the surface features of the planet. USGS scientists and engineers are investigating new types of satellite systems and sensors, studying promising new data sources, developing new data acquisition programs and sources, and assessing the potential for new data applications. In addition, the USGS is seeking new ways to make remotely sensed data products more accessible to users, as well as ways to expand and enhance overall use of remotely sensed data and remote-sensing technology.

Remotely sensed data are complemented by a wide range of on-the-ground measuring and monitoring systems that provide a higher level of detail and accuracy. Many of these systems are described elsewhere in this book, including the streamgaging network, the Advanced National Seismic

System, and the North American Breeding Bird Survey. The Earth and its processes are always more complex than we can imagine. As a result, long-term, geographically diverse data sets reveal much more than expected and are essential tools for protecting lives and property; managing water, energy, mineral, and biological resources; and enhancing the quality of life. New monitoring—whether space-based, airborne, or on the ground—leads to advances in our understanding of the Earth and its processes.

Real-Time Data—Critical for Many Users

In the past 25 years, technological advances have dramatically improved our ability to communicate information quickly and accurately to an extremely diverse group of

Jean Parcher

Jean Parcher can trace her interest in geography to her grandfather, a worldwide shipping merchant. He inspired a wanderlust that has taken her on many journeys through Central and South America, including a 3-year tour of duty with the Peace Corps, where she volunteered in rural community development for the Costa Rica Bureau of Indian Affairs. Her experiences there provided her with an excellent foundation to succeed in her role of coordinating the acquisition of geospatial mapping data in Texas. The cost of generating quality sources of mapping data is expensive unless the burden is shared by many partners and the data made publicly available. Parcher coordinates partnerships by bringing together various State, Federal, and Mexican agencies with similar geospatial data mapping needs, such as economic development planning, homeland security planning, natural hazards analysis, agricultural production, and monitoring urban growth changes to the environment.

To better understand the needs of her partners, Parcher wanted to broaden her knowledge in science issues. Through the USGS graduate program, she found the opportunity to return to her studies. Her research on comparing land use and flood hazards for two cities along the U.S./Mexico border was selected as one of the best theses at the University of Texas in 2003. Currently, she is working with the three other USGS disciplines to develop a bi-national geographic information database along the U.S./Mexico border to analyze environmental affects on human health issues. Her ability to conduct meetings in Spanish and to cross cultural barriers, honed during her Peace Corps days, greatly enhances her work and contributes to the success of the project.

1980

Mount St. Helens erupts on May 18. In 1978, USGS geologists had forecast that a violent eruption was probable. On March 27, it is determined that the possibility of eruption is great enough to issue a hazard warning. The eruption on May 18 results in 57 deaths and flooding on the Toutle and lower Cowlitz Rivers.

1980s

Instream Flow Incremental Methodology (IFIM) is developed to provide a mechanism for managers to determine the effects of their decisions on riverine systems. This integrates several scientific disciplines into a single comprehensive habitat-analysis tool.

Web Accessibility

Anything from a game show question to a flood or earthquake can cause hits on the USGS Web site to jump significantly. Whether it's idle curiosity about minerals, an unquenchable thirst for knowledge about water, or an urgent need for information about a hazard—the USGS is working to make the information available to anyone who may want or need it. The USGS ensures that all breaking news published on the Web is accessible to adaptive technology. Its top 20 Web sites all comply with the requirements of the Rehabilitation Act of 1973, Amendments of 1998, Section 508. Procedures are in place on all USGS Web pages to provide alternative access for applications that are inherently not accessible to adaptive technology. The USGS has collaborated with other Federal agencies and industry on how to provide alternate access to its maps. In the early 1970s, the USGS was one of the first Federal agencies to produce Braille and textile maps. Many USGS maps are so complex that no technology is currently available to make them accessible to people who are visually impaired. To address their needs, USGS has adopted several strategies, including links to descriptive text for the less complicated maps and contact information on all USGS Web pages so that people can call or e-mail their questions and ask for assistance in finding information. The USGS is firmly committed to ensuring that its scientific research results, in all formats, are fully accessible to everyone.

users. From real-time stream-flow monitoring to mapping wildfire perimeters, providing earthquake information minutes after an event, and enabling access to more than 100 years of published books and maps, the World Wide Web is a powerful tool for tapping into the wealth of USGS information. The USGS is committed to ensuring that all electronic and information technology developed, procured, maintained, or used by the USGS is accessible to people with disabilities, including both employees and the customers we serve.

The USGS operates a nationwide network of nearly 7,000 real-time streamgages and more than 1,000 real-time water-quality monitoring stations that provide critical hydrologic information in time for effective decisionmaking about flooding, drought, and water supply. The network is funded by the USGS and more than 800 other Federal, State, and local partners who use the information for planning and management of water resources; design of bridges, dams, and treatment facilities; research on changes in land and water use; and safe recreation and enjoyment on rivers, lakes, and streams. The data from the streamgages are transmitted by satellite and radio to a distributed system of mirrored servers. In Louisiana, which leads the Nation in property damage caused by floods, an innovative interactive flood-tracking chart for the Amite River basin is helping to protect lives and property. People in the Amite River basin can click on a map of the basin and place the cursor over any of the nine highlighted gages to see a chart showing current river level, the National Weather Service flood stage, and the highest recorded river levels at that location. Using this information, they can make informed decisions about evacuation and protection of personal property.

In much of the West, fire is a primary concern. During the 2000 fire season, over 79,000 wildfires burned nearly 7 million acres of land along with hundreds of structures and valuable natural resources. In response, the USGS partnered with other Federal land man-

1980s

Research at the USGS Patuxent Wildlife Research Center, Md., and the National Wildlife Health Center, Wis., leads to the eventual banning in 1991 of lead shot for waterfowl hunting and the acceptance of non-toxic alternatives. USGS scientists also document toxic effects of selenium on birds and other biota at the Kesterson Wildlife Refuge and study many other sites in the Western United States that received irrigation drainwater.

1981–1993

Dallas Lynn Peck is the 11th Director of the USGS.

1982

Part of the USGS staff and operating budget is split off to become the Minerals Management Service. The Conservation Division and part of the marine geology program, about 23 percent of the personnel and 29 percent of the operating budget of the USGS, are reassigned to the new agency.

1983

President Reagan proclaims an Exclusive Economic Zone (EEZ), which extends jurisdiction of the United States for a distance of 200 nautical miles seaward of the Nation's shorelines. Geological Long-Range Inclined Asdic (GLORIA) imagery is the first systematic mapping of sea-floor morphology and features in the deep ocean.

agement agencies to create an online resource for fire managers. This Web-based mapping tool allows fire personnel and the public to access online maps of multiacre blazes. Fire locations, weather data, and near-real-time satellite data are integrated with shaded relief, roads, cities, and other information. Fire perimeter data are updated daily with input from incident intelligence sources, GPS data, and infrared imagery from airplanes and satellites. The GeoMAC (Geospatial Multi-Agency Coordination) Web site allows users anywhere in the world to manipulate map information displays, zoom in and out to display fire information at various scales and detail, and download and print material for use in fire information and media briefings, dispatch offices, and coordination centers.

The USGS has Federal responsibility for monitoring

and reporting on earthquake activity across the United States and internationally, and this activity increasingly takes place through the Web. The media, emergency responders, public officials, and other individuals around the world now depend on USGS Web-based earthquake reports as the primary source for real-time information about destructive earthquakes. Accordingly, Web servers that provide earthquake information see tremendous surges in traffic after an earthquake. The typical pattern shows an abrupt increase in the rate of hits to as much as 1,000 times normal

usage, peaking between 3 and 30 minutes after the event. To accommodate this surge in traffic, USGS contracted with Akamai Technologies to provide Web content delivery that is geographically distributed around the world and spans a large number of networks, thus avoiding the possibility of a single or regional point of failure, as well as offering enhanced performance to a distributed base of USGS customers. With this service, USGS can guarantee rapid delivery of the information that emergency managers and the public need after a significant earthquake.

Thomas Loveland

Thomas Loveland

As a South Dakota native, Tom Loveland has never had to venture far to know what's going on around the world. He views the Earth from a vantage point in the farmland of eastern South Dakota, the USGS EROS Data Center, where images of the world stream in from Earth-orbiting satellites. For nearly 25 years, 15 of which have been with the USGS, Loveland has used remote sensing to study the Earth's changes.

Loveland, a USGS research geographer, uses images of the Earth to study the changing patterns of land use and land cover. He's learned that changes in the land can lead to unexpected changes in the weather, water quality, and wildlife habitat and can play an important role in stimulating local economies. He led a team that created the first detailed and validated global map of land use and land cover. Working with geographers from USGS facilities across the country (both in the field and with satellite images in laboratories), Loveland is striving to provide the most detailed documentation ever assembled on the rates and causes of land-use and land-cover change in the United States.

Over the years, the scale of Loveland's investigations has shifted from global to national to regional. He has been involved in design studies for monitoring landscape conditions in more than 25 countries, including such remote corners of the world as Somalia, Uganda, and Chad. He has also studied land-use and land-cover characteristics in settings from Alaska to Florida and most places in between.

Loveland is active in the development of national and international land-cover science initiatives and has held many leadership roles. He has been honored with career achievement awards from the American Society of Photogrammetry and Remote Sensing and the Association of American Geographers.

1984

The first National Water Summary is published, describing hydrologic events and water conditions for the water year. Also, the first six studies of major regional aquifer systems studies are published.

1986

The National Water-Quality Assessment (NAWQA) pilot program begins.

1986

The National Geographic Names Database is created.

Feature Name	St	County Equivalent Name	Type	Latitude nn°nn'nn"	Longitude nnn°nn'nn"	USGS 7.5' Map
Anniversary Park East	MA	Middlesex	park	422204N	0711750W	Natick
Anniversary Mine	NM	Luna	mine	320423N	1073652W	Gym Peak
Anniversary Mine	NV	Clark	mine	361249N	1144234W	Callville Bay
Anniversary Hall	OH	Gallia	building	385255N	0822245W	Rio Grande
Anniversary Arch	UT	Grand	arch	384700N	1094036W	Klondike Bluffs

David J. Wald

Before 1999, the only information available in the minutes following a damaging earthquake was the location and magnitude of the event, which simply was not enough to help emergency responders determine where their help was needed most. Dave Wald, a research seismologist at the USGS, revolutionized earthquake response with much-needed, real-time seismology tools that use the speed and accessibility of the Web to deliver post-earthquake information when it's needed most—right after an earthquake. One such tool is ShakeMap. ShakeMap uses shaking levels recorded by a dense network of seismic stations in the Advanced National Seismic System to map the distribution of shaking, pointing to areas most likely to have experienced damage. These maps, available online within 5 to 10 minutes after an earthquake, provide the basis for emergency response coordination, estimation of damage and losses, and information for the public and the media. ShakeMap minimizes business disruption, reduces the economic impact of an earthquake, and enables a more timely and prioritized response that helps save lives. ShakeMap is now running in five metropolitan areas across the United States, and implementation efforts are underway in a number of other urban centers.

Another Walk contribution, the Community Internet Intensity or "Did You Feel It" maps, uses citizens' observations to collect information about ground shaking. Immediately after shaking subsides, Web users can log onto a Web site and record what they felt and observed during an earthquake. "Did You Feel It" has become a popular tool for the public, with more than 350,000 individual entries collected for earthquakes nationwide. Wald's maps take advantage of the vast numbers of Web users to provide a more complete and rapid description of what people experienced, the effects of the earthquake, and the extent of damage than was possible with traditional ways of gathering information.

Earthquake Monitoring for a Safer America

To collect the essential earthquake information that people need and to address the sharply increasing cost of earthquakes in the past two decades, the USGS is developing the Advanced National Seismic System (ANSS), a nationwide network of at least 6,000 state-of-the-art earthquake sensors, or seismometers, on the

ground and in buildings in 26 urban areas across the United States that face significant risk from earthquakes. These digital instruments record how different types of ground and structures shake during a strong earthquake, enabling engineers to greatly improve the current design standards for buildings, lifelines, and other structures and to identify critical weaknesses of current structures.

ANSS is based on the success of TriNet, a 5-year collaborative project between the USGS, the California Institute of Technology, and the California Geological Survey, with funding from FEMA. That partnership began in 1997 with the goal of creating a better real-time earthquake information system for Southern California. Although still under construction, ANSS is already providing detailed information about the severity of earthquakes to emergency responders in government

and the private sector, helping to guide the response and speed the economic recovery. One key ANSS product is ShakeMap, an online map of the extent of ground shaking that can be posted on the Web within 5 to 10 minutes of an earthquake. First developed in Southern California as part of TriNet, ShakeMap is a valuable tool for emergency response, loss estimation, and public information.

Earthquake monitoring instruments are also being installed in a dense network around and underneath the town of Parkfield in central California. Moderate earthquakes have occurred on the Parkfield section of the San Andreas fault at fairly regular intervals for the past 150 years. The USGS and the California Geological Survey began closely monitoring the area in 1985. The latest step is a project, the San Andreas Fault Observatory at Depth, funded primarily by the

1988

The Alaska Volcano Observatory begins as a joint program of the USGS, the Geophysical Institute of the University of Alaska Fairbanks, and the Alaska Division of Geological and Geophysical Surveys.

1988

In cooperation with the Bureau of Census, the USGS completes a 1:100,000-scale digital database of the transportation and hydrology features of the conterminous United States.

1988

The USGS joins the commercial Internet at metropolitan area exchanges (MAEs) on the East and West Coasts.

National Ice Core Laboratory

Locked away in tiny bubbles, trapped beneath layer upon layer of ice, are the secrets of the climate and atmosphere of the ancient Earth. By extracting samples of ice from glaciers, scientists at the USGS National Ice Core Laboratory (NICL) in Denver, Colo., are learning what the Earth was like hundreds of thousands of years ago. They are gathering information on temperature, precipitation rates, chemistry and gas compositions of the lower atmosphere, volcanic eruptions, solar variability, sea-surface productivity, and other climate and atmospheric indicators. An area of increasing interest is the biological history that may be discovered in ice cores, including microbes, DNA, pollen, and the exciting possibility of life in subglacial Antarctic lakes.

Ice comes to the laboratory in the form of "cores"—cylinders drilled from the polar ice sheets with drills that are like large hollow screws. The cores are transported from Antarctica and Greenland to the lab in Denver, where the staff assists scientists from around the world in uncovering the information locked inside. Safely transporting the cores to Denver requires the cooperation of many organizations, including the National Science Foundation, polar logistics contractors, and the New York Air National Guard. Once the cores have arrived at the lab, the precious collection (valued at perhaps $100 million) is preserved at a constant –36° C in the main storage area.

The outreach efforts of the lab are extensive. More than 1,500 members of the scientifically interested public visit each year, media representatives tour the lab and have made documentary films about the work conducted there, and a rapidly growing database about the ice core holdings is available on the NICL Web site <*nicl.usgs.gov*>. This database is an expanding and evolving record of the inventory and its history; its aim is to create a digital photographic record of the core holdings available over the Web. At the NICL, studying ice is not just cool—it's the key to understanding how our planet and its climate have changed over hundreds of thousands of years.

Robert I. Tilling

Bob Tilling is no stranger to the hazards of our restless planet. He worked as a USGS volcanologist for more than 30 years and has served as Scientist-in-Charge at the USGS Hawaiian Volcano Observatory, as Chief of the Office of Geochemistry and Geophysics during the catastrophic eruption of Mount St. Helens, and as Chief Scientist of the USGS Volcano Hazards Team.

One of Tilling's greatest contributions has been his dedication to communicating his knowledge to others. Over the years, Tilling's publications have been some of the most popular in USGS history. He took the lead in communicating scientific goals and applications to the public before concepts like "outreach" and "public service" became widely recognized as critical to USGS missions.

Tilling has written many technical and general-interest articles, including co-authoring the popular USGS book on plate tectonics, "This Dynamic Earth," with Jackie Kious. This publication set USGS records for hits on its Web-based version; in 2003, there were close to 2.7 million inquiries to the site. "This Dynamic Planet," an educational map that Tilling coauthored, has sold more than 80,000 copies and is the most requested map in the history of the USGS. His scientific publications, his work in training of foreign volcanologists, and his talents both as scientist and administrator have significantly increased the knowledge and awareness of volcanic hazards both in the scientific community and among the public.

1989
Loma Prieta, Calif., earthquake—magnitude 6.9.

1990

The Federal Geographic Data Committee (FGDC) is established.

1990s
USGS research about the ages and depositional history of mineralized deposits identifies volcanogenic massive sulfide deposits, the source of base and precious metals. Mineral exploration and land use planning decisions about land near Wrangell, Alaska, benefit from this research.

Thomas S. Ahlbrandt

These days, oil and gas resources are a very hot topic. Thomas S. Ahlbrandt's in-depth knowledge and diverse experience have made him highly sought after for scientific committees and technical review panels. He is frequently invited to share his expertise with the news media, domestic and foreign government agencies, academic institutions, and industry.

Ahlbrandt has conducted and directed research investigations on the origin, distribution, quantity, and quality of oil and gas resources of the United States and the world. After joining the USGS, Ahlbrandt helped to establish the importance of the USGS energy research program to meeting national energy needs and undertook a major investigation of the oil and gas resources of the United States. The study provided the most thorough, best documented, and most scientific assessment of oil and gas resources ever published and has been enthusiastically received by industry, State Governments, and Federal agencies.

As Chief of the World Energy project, Ahlbrandt led efforts to conduct collaborative, impartial, and integrated geologic studies that provided an understanding of the occurrence, quantity, and quality of world energy resources.

Ahlbrandt's research in the field of eolian (wind-deposited) sediments has also brought him worldwide recognition. He developed the eolian sand-movement chronology for the Great Plains for the past 10,000 years. This landmark study is of great importance for assessing the role of the greenhouse effect on the present climate of the Great Plains.

National Science Foundation (NSF) and jointly led by USGS and Stanford University scientists, to drill a deep hole through the San Andreas fault. A pilot hole more than a mile deep was drilled next to the deep drill-hole site in 2002. Monitoring instruments will be installed in the pilot hole and the deep drill hole to directly reveal, for the first time, the physical and chemical processes controlling earthquake generation on an active fault. This drilling project is just one component of NSF's EarthScope program. The other elements are an array of seismic monitoring instruments that is coordinated with ANSS and a network of deformation sensors for the Western United States. The cooperative efforts of the USGS, NSF, and many other scientific agencies and institutions will help achieve a better understanding of what happens on and near faults during the earthquake cycle.

Partnerships for Collecting Geographic Data

Geographic data are vital for homeland security, response to natural disasters, efficient delivery of government services (such as the decennial census of population), scientific investigations, and effective land and resource management. In early 2001, the USGS announced a new vision for providing geospatial information for the Nation. This new vision, called *The National Map <www.nationalmap.gov>*, builds on John Wesley Powell's use of the phrase in his early testimony to Congress regarding topographic mapping and is the transformation of the USGS's historic topographic mapping mission for the 21st century.

Partnerships are the foundation of *The National Map*. The role of the USGS in national mapping has changed dramatically from that of a first-line producer of data and maps to that of a community organizer, partnership facilitator, standards coordinator, and data integrator. State and local governments have extensive mapping capabilities and workforces in the field who are in direct touch with changes on the ground. The private sector provides national information for important data themes and sources from which updated information can be extracted. Two partnerships demonstrate many of the concepts and future capabilities of *The National Map*. The State of Delaware's DataMIL brings together local, State, and Federal data and

1991
The National Oil Shale Database is created.

1991
The Southern California Earthquake Center is established in Pasadena, Calif., with funding from the National Science Foundation and USGS.

1991
1:24,000 topographic quadrangles are completed for the conterminous United States: More than 33 million person hours and $1.6 billion were invested for this 7.5-minute series of USGS topographic maps that began in the 1930s.

1992
The National Cooperative Geologic Mapping Program is established.

demonstrates how maps made to user specifications can be derived from these data. This site, at *<http://datamil.udel.edu/>*, also provides users with a means to contribute updated information to data stewards for incorporation in the component datasets. At *<http://www.tnris.state.tx.us/stratmap/index.htm>*, the Texas Natural Resources Inventory System makes integrated digital data and derivative maps available.

The USGS recognized early on that the technological revolution in computing (GIS, microcomputers) and telecommunications (GPS, the Web) would forever change the way in which the agency accomplished its mission. Senior leaders understood the importance of moving quickly to bring together other Federal agencies to develop standards, protocols and new approaches.

For example, the USGS developed the revolutionary conceptual framework that came to be called the National Spatial Data Infrastructure, convened the first interagency meetings on geographic information systems and data sharing in the mid-1980s, and has provided an institutional home for the Federal Geographic Data Committee secretariat since 1990. The USGS led the development, adoption, and implementation of the content standard for digital geospatial metadata.

The USGS has also led in developing telecommunication protocols essential for search and retrieval of distributed geospatial data and content standards for geographic data themes. Coordinated geographic data collections enhance the quality of our lives every day. We depend upon seamless delivery of an array of municipal services (school bus routing, traffic analysis, sanitation systems, and so on) and rapid response to natural disasters, industrial accidents, and homeland security alerts. Because geographic data are essential to these services, our society depends upon accurate, current, and regionally consistent geospatial data.

GIS Technology

Almost all major issues confronting society in the 21st century have a spatial component—they occur somewhere. Urban growth, water pollution, invasive species, coastal erosion, nuclear waste disposal, and hazard mitigation are just a few of the complex issues facing the global community. Geographic information systems provide a technology and method to analyze these spatial issues on, above, and below the surface of the Earth.

National Wetlands Research Center

Take some coastal storms and occasional hurricanes, mix in a dash of sinking wetlands and rising sea level, add decades of human alteration of the coastal environment and several million people; stir well to combine. The result: a perfect recipe for a disappearing coastal landscape. But by adding focused science and strong partnerships to the recipe, USGS is cooking up a solution to the problem of Louisiana's vanishing coast.

Louisiana's coastal ecosystem has been eroding into the Gulf of Mexico at an average rate of 34 square miles a year for the last 50 years. Since 1932, 1,900 square miles of coastal land have been lost, and as much as another 700 square miles may disappear in the next 50 years if nothing is done. This projected loss is more land than the State of Delaware plus the Washington, D.C., and Baltimore, Md., metropolitan areas.

Coastal land loss was first identified as a concern for the Gulf of Mexico Coast in 1970 by scientists now working for the USGS; today it is clear that the problem is national, not local, in scope. Coastal erosion affects all 30 coastal States and all the U.S. island territories. If erosion continues at this rate, the losses to wildlife, the economy, the Nation's oil and gas infrastructure including its strategic reserves, and human property and safety will be staggering.

Restoring the coast may take billions of dollars, and the decisions regarding restoration must be supported by objective science. USGS scientists at the National Wetlands Research Center in Lafayette, La. are focusing their biological, geological, hydrological, and geographical expertise to address this critical issue in an integrated and solution-oriented approach, in partnership with State and other Federal agencies, universities, and nongovernmental organizations.

1993
The National Biological Survey is created to consolidate the biological research functions of many DOI bureaus, including the National Park Service, the Fish and Wildlife Service, and the Bureau of Land Management. The creation of an independent scientific agency insulates scientific research from those who manage lands and draft government regulations.

1993
Excessive rainfall in the Midwest during much of the spring and summer leads to significant flooding along the Mississippi River basin; damages are estimated at $20 billion, making this the costliest flood in U.S. history.

1994–1997
Gordon P. Eaton is the 12th Director of the USGS.

Roger L. Payne

The United States is one of the largest countries in the world—keeping track of all of its geographic names is a very large task. For more than 30 years, Roger L. Payne has been involved with all aspects of the research and standardization of geographic names. Along the way, he has made quite a name for himself, as well.

Payne is the Executive Secretary of the U.S. Board on Geographic Names, an interdepartmental body that standardizes geographic name usage throughout the Federal Government. He directs both the Federal and USGS geographic names programs. He is the architect of the Nation's official automated geographic names repository, the Geographic Names Information System (GNIS). He designed the extensive collection phase to populate the database and created an automated program to enlist Federal, State, and local participation to help maintain it. He gave the GNIS its Web presence <***geonames.usgs.gov***>, allowing people to interactively submit their information and requests to the GNIS and the names board.

Among his many distinguished positions, Payne has represented the Nation, the names board, and the USGS in all matters of toponymy (the study of geographic names) at the United Nations since 1987, and he has been head of the U.S. delegation since 1993. He's participated in exchange programs with Antarctica, China, and the Historically Black Colleges and Universities program. He was instrumental in the design of the Pan American Institute of Geography and History's (PAIGH) toponymic program and has served as adjunct faculty at three universities. Throughout his career, Payne has been dedicated to helping people better understand what's in a name.

The Earth's climate, natural hazards, population, geology, vegetation, soils, land use, and other themes can be analyzed in a GIS by using computerized maps, aerial photographs, satellite images, databases, and graphs. By analyzing themes about the Earth's hydrosphere, lithosphere, atmosphere, and biosphere, a GIS helps people understand patterns, linkages, and trends about our planet.

Over the past 30 years, GIS has quietly transformed decisionmaking in universities, government, and industry by bringing digital spatial data sets and geographic analysis to desktop computers. It has developed into geographic information sciences, which includes geographic information systems as well as the disciplines of geography (examining the patterns of the Earth's people and physical environment), cartography (mapmaking), geodesy (the science of measuring and surveying the Earth), and remote sensing (studying the Earth from space). A GIS not only manages location-based information, it also provides tools to visualize, query, and overlay spatial information in ways not possible with traditional methods – whether the subject is characteristics of a watershed, bird population distribution, or areal extent of mineral deposits.

A GIS is more than just "computer mapping." Through GIS, databases can be linked to maps to create dynamic displays. More importantly, GIS provides tools to analyze and understand these databases in ways not possible with traditional tabular presentations, such as spreadsheets. GIS and associated technologies are essential to the USGS both in performing our scientific investigations and in delivering the results of our science to our customers and partners.

Since the early 1980s, the USGS has been using some form of GIS technology to collect and analyze geospatially referenced data. In 1983, USGS scientist Charles Robinove and his colleagues published the first scientific analysis using GIS, for the Fox-Wolf River basin in Wisconsin. The USGS will increasingly be asked to provide information needed to address overarching environmental problems. Solutions to

1994
The USGS joins the World Wide Web.

1994
A Presidential Executive Order calls for the establishment of the National Spatial Data Infrastructure (NSDI). The Federal Geographic Data Committee launches the NSDI Cooperative Agreements Program, a major grant program to stimulate NSDI activities nationally.

these problems will require a broad and integrated research program that is capable of developing highly complex system models, using geospatially referenced data.

To meet these challenges, the USGS will continue to develop a more integrated bureauwide "Enterprise GIS" capability that can support more effective collection, application, sharing, and reuse of GIS analysis tools and GIS data across the entire organization. New GIS tools for data sharing, data merging, data revision, data analysis, and data visualization will also be developed. One key example is mobile computing—using hand-held, wireless devices for GIS work (including GIS data collection and analysis in real time right in the field). This technology can substantially reduce the time it takes for scientists to collect and interpret data to inform decisionmaking.

Understanding Landscape Change

The Earth's surface is changing rapidly and at an increasing rate, at local, regional, national, and even global scales. Some of these changes have natural causes, such as landslides or drought; others, such as resource extraction, urban growth, and water-resource development, are examples of human-induced change that have significant impact on people, the economy, and resource availability. Improved understanding and information about the consequences of landscape change are needed to help decision-makers in land-use planning, land management, and natural resource conservation. One example where USGS landscape-change science is making a difference is the study of urban growth and land-use

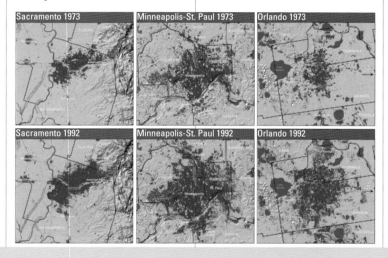

Sacramento 1973 · Minneapolis-St. Paul 1973 · Orlando 1973
Sacramento 1992 · Minneapolis-St. Paul 1992 · Orlando 1992

1994
Northridge, Calif., earthquake—magnitude 6.7.

1995
The National Biological Survey changes its name to the National Biological Service.

1996
In Congressional appropriations process, the National Biological Service is merged with the USGS to create a single DOI science agency, and the USGS Biological Resources Division is established.

1996
The Bureau of Mines is abolished, and the minerals information team and other programs are transferred to the USGS.

Arlen W. Harbaugh

Arlen W. Harbaugh is one of the most sought-after men in science. His advice and counsel are sought by a wide range of people, from the newest USGS employees to the most experienced quantitative hydrologists. He is not only one of the world's most renowned ground-water hydrologists; his outstanding ability to understand complex hydrologic problems is combined with an unselfish attitude toward helping others.

Harbaugh began his career in Long Island, where he pioneered an innovative hybrid approach to ground-water modeling. It was a computerized data acquisition system for one of the largest three-dimensional electric-analog ground-water models in existence and led to one of the first digital three-dimensional ground-water-flow models.

While in New Jersey in the late 1970s and early 1980s, Harbaugh worked with Michael McDonald to develop MODFLOW, a numerical ground-water flow model. MODFLOW has proved to be an outstanding contribution to the field of hydrology and has influenced ground-water studies worldwide. It is now the standard tool used by governmental agencies and the private sector to help protect and manage ground water and has become the most widely used ground-water flow model in the world.

Harbaugh's achievements have earned him a special recognition award from the National Ground Water Association, the USGS Meritorious Service Award, and the James R. Balsley Award for Technology Excellence.

Flagstaff Science Center

The Flagstaff Field Center was established in 1963 to map the Moon and help train astronauts destined for its surface. A wide variety of people, including scientists, cartographers, computer scientists, administrative staff, students, contractors, and volunteers, work there in support of efforts to explore our solar system. Throughout the years, the program has helped process and analyze data from missions to different planetary bodies, determine potential landing sites for exploration vehicles, map our neighboring planets and their moons, and conduct research to better understand the origins, evolutions, and geologic processes operating on these bodies.

The first group of USGS scientists to find a home in Flagstaff was not the astrogeologists, but rather hydrologists. For 55 years, USGS has been installing, monitoring, and maintaining streamgages in northern Arizona, and scientists at Flagstaff continue to work with the cities, counties, and Tribal communities of the region to study the availability and sustainability of surface- and ground-water resources. More recently, the mission of the Flagstaff Science Center has expanded to include research on biological issues facing the Southwest. Biologists at the Southwest Biological Science Center at Flagstaff study issues such as how ancient vegetation on the Colorado Plateau has responded to climate change and how water released from Glen Canyon Dam affects ecosystems and habitats along the Colorado River from Glen Canyon Dam through the Grand Canyon to Lake Mead.

Arizona is a long way from Hawaii, but geographers at Flagstaff are using remotely sensed images to map coral reefs off the island of Molokai and to determine how land-use patterns affect erosion, which can dump damaging amounts of sediment onto coral reefs. Closer to home, researchers are developing methods to forecast dust emissions in the Southwestern United States and to evaluate the possibility of disease outbreaks caused by wind-blown dust.

change. By comparing remotely sensed images of the same region taken in the 1970s, 1980s, and 1990s, decisionmakers gain insight into both the causes and the consequences of urbanization and anticipate likely future patterns of urban growth.

Suburbanization has been an increasing trend since World War II. By the end of the 1960s, suburbs were no longer simply bedroom communities for commuting city workers but were emerging as focal points for the retail and service sectors of the urban economy. The paired images show that suburbs often developed along main highways in a linear fashion, linking pre-existing outlying communities with the main city. Some cities continue to grow significantly, while others have reached certain physical limits of geography that restrict their growth. For example, the Los Angeles basin and San Francisco Bay regions have little remaining land available for development. As a consequence, development there is spilling out into the Mojave Desert beyond the San Gabriel Mountains or down into valleys in Contra Costa, Alameda, and Santa Clara Counties. The dynamics of American cities have not stopped; only their future evolution is yet to be determined. USGS scientists will continue to provide information on the status and trends in our changing landscape to inform regional and national decisions.

1997

USGS adopts a new identifier and the motto "science for a changing world."

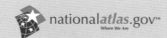

1997

The second National Atlas (electronic version) begins.

1997

The Johnston Ridge Observatory is dedicated in memory of the USGS scientist who died on Mount St. Helens.

USDA, Forest Service

America's Least Wanted: Invasive Species

The introduction of non-native species into the United States is increasing rapidly. Some, such as pets and ornamental plants, are introduced intentionally. Others arrive as hitchhikers associated with global travel and trade. Based on past experience, 10 to 15 percent of these nonnative species can be expected to establish free-living populations, and about 1 percent can be expected to become invasive and cause significant negative impacts to ecosystems, native species, economic productivity, or human health. Notorious examples include West Nile virus; exotic plants like tamarisk, cheatgrass, and kudzu; and animals such as zebra mussels, fire ants, and the brown tree snake—all of which have inflicted costly damage across the continental United States and the Pacific Islands.

The spread of invasive species is considered one of the most serious ecological problems facing our Nation in the 21st century, second only to habitat destruction. The invaders include plants, animals, and microbes that have been introduced into an environment where they did not evolve and often have no natural enemies to limit their reproduction and spread. Many invaders cause huge losses in agriculture, livestock, fisheries, and other resource production

systems. Some significantly alter ecosystems, resulting in costly damages due to increases in fire, flooding, and erosion. A few are vectors, or carriers, of human diseases. Unfortunately, invaders often go unnoticed until they have spread to many locations, making eradication difficult. Early detection and rapid response to invasions are essential if we are to prevent major damage to the health of our ecosystems and economy—an ounce of prevention is truly worth a pound of cure.

1998–Present
Charles G. Groat is the 13th Director of the USGS.

1999
Landsat 7 is launched.

1999
West Nile virus is detected in the United States in New York State.

1999
The National Park Service begins its Natural Resource Challenge Program. Over time, it will greatly increase attention to biology in terms of inventory, monitoring, vegetation mapping, removal of exotic plants, and support for partners to conduct research in parks.

June M. Thormodsgard

There is a subtle yet important difference between being a good leader and being a good manager. June M. Thormodsgard is good at both. As a leader in the scientific field of remote sensing, she has been involved in the strategic planning and implementation of a number of critical programs in the USGS. As a manager, she has promoted a productive, collegial work environment for scientists, engineers, and operational staff. She has motivated and inspired teams of scientists who are, in their own right, national leaders in scientific research.

Thormodsgard has led teams of multidisciplinary researchers whose mission is to help earth scientists and resource managers use satellite images. These teams serve as a bridge between academia, government, private research institutions, and practical programs involved in the mapping, monitoring, modeling, and management of earth resources. Together, she and her staff help modern society use satellite data to deal with environmental and resource inventory challenges.

She has served as the DOI Coordinator for LANDFIRE, a joint project with the U.S. Forest Service to map fire fuels with satellite imagery and biophysical data. For this project, 32 separate organizations and projects related to fire science and management were contacted to promote collaboration and reduce overlap.

Thormodsgard has worked to personally mentor the next generation of leaders, by forming and leading a Women Helping Women group. The group meets regularly to assist women in their efforts to improve their education, advance in their jobs, and handle the unique pressures associated with balancing work and home responsibilities.

Mary Lou Zoback

As a geologist and geophysicist for the USGS, Mary Lou Zoback works to figure out what on Earth—or rather, what in the Earth—causes earthquakes. Specifically, she is interested in the origin of stresses in the Earth's crust and their relationship to earthquakes, both along plate boundaries like the San Andreas fault and within plate interiors. Working with her husband and colleague Mark Zoback, of the Geophysics Department at Stanford University, she developed methods for determining the present-day distribution of forces acting on the Earth's thin outermost layer, or crust, from a variety of data and observations.

From 1986 to 1992, Zoback led the World Stress Map project, an international collaboration of 40 scientists from 30 different countries, which compiled and interpreted a global database on present-day stresses in the Earth's crust. This work demonstrated that broad regions in the interior of most tectonic plates are subjected to relatively uniform stresses that result from the same forces that cause plate motion.

Zoback has received numerous awards and served on national and international scientific review and advisory panels. She was one of 17 scientific experts selected by the National Research Council to evaluate the potential for ground water to rise to the level of the proposed high-level radioactive waste repository proposed for Yucca Mountain, Nev., and currently serves on the Board on Radioactive Waste Management. She is a member of the National Academy of Sciences.

Invasive species are a growing threat to the Department of the Interior's stewardship of the Nation's natural resources. Invasive plants are currently estimated to infest more than 2.6 million acres in the National Park System. By various estimates, these invasive species contribute to the decline of 35 to 46 percent of U.S. endangered and threatened species. In some regions, entire ecosystems are being transformed. In the Intermountain West, invasive plants such as cheatgrass are increasing the frequency and intensity of fire, replacing native species, and damaging rangeland values on our public lands. Aquatic invaders, such as the zebra mussel, Asian carp, giant salvinia, and purple loosestrife, are transforming our wetlands and inland waters and reducing their value for recreation and wildlife.

To help address the challenge, the USGS has established the Institute of Invasive Species Science, a consortium of partners including USGS research centers, NASA, other government agencies, nongovernmental organizations, and universities. The vision of the institute is to provide national leadership in invasive species science and to work with others to disseminate and synthesize current and accurate data and research from many sources to predict and reduce the effects of harmful nonnative plants, animals, and diseases in natural areas and throughout the Nation.

Science: the Foundation for Sound Decisions

The scientific mission of the USGS, its national perspective, and its nonregulatory role enable the USGS to provide information and understanding that are policy relevant but policy neutral. That independence from policy and regulation has been essential to enable the USGS to provide the unbiased science needed to make informed resource management decisions. In the last few years, the USGS has set in place a number of mechanisms to better coordinate with the other DOI bureaus and has dedicated funds to address specific departmental science needs. The goal is to enhance science support to meet priority needs for bureaus whose responsibilities include the management of the Nation's land and natural resources. Particular emphasis has been placed on restoring degraded habitats, wildlife issues, hydrologic and biologic issues related to energy development, and ground-water availability studies. Long-range planning

1999
The Geographic Information Office (GIO) is established.

1999
The USGS implements an enterprise dial-in remote access, connecting employees from multiple locations, including their homes.

1999
Hurricane Floyd causes major floods along the East Coast.

2000
The USGS determines how the genetic makeup of Atlantic salmon differs across their North Atlantic range. The discovery is instrumental in providing protection and management strategies for the recovery of this fish in its spawning rivers in Maine.

NPS

NPS

NPS

and priority setting occur at DOI headquarters, but the more significant interactions of USGS scientists and DOI land and resource managers occur at the local level, managed in the regional structure of both the USGS and the other bureaus.

In the USGS Eastern Region, science activities focus on environmental and human impacts on ecosystems. Impacts of sea-level rise on low-elevation coastal habitats, effects of urban dynamics on the sustainability of inner city public lands, and ecosystem sustainability studies on impacts of altered freshwater flow on

bay and coral reef health are critical issues for both the Fish and Wildlife Service and the National Park Service.

The USGS Central Region has placed a high priority on invasive species in riparian and aquatic habitats; restoration ecology to assist other bureaus in developing interdisciplinary strategies to restore habitats affected by fire, abandoned mines, or energy extraction; coal-bed methane production on public lands; and assessment and modeling of ground-water resources at regional scales.

In the USGS Western Region, science activities

focus on integrated issues in key areas of Alaska, the Great Basin, the arid West, and the Pacific Islands. Projects on the North Slope of Alaska focus on energy development-related physical and biological issues to provide science information for sound management decisions. In the Great Basin, ecosystem sustainability depends on results of interactions of wildfire and invasive species, as well as on impacts of geothermal development. Water continues to be a critical issue in the arid West, where there is a need to develop a clearer understanding of instream needs for aquatic

Robert H. Meade

Bob Meade's career with the USGS spans more than 50 years; his work has given society a better understanding of how sediment moves through rivers and estuaries, thereby reshaping the landscape. His contributions have withstood the test of time and become accepted concepts in textbooks throughout the world. Meade has extended our knowledge of land subsidence; proposed and substantiated the concept that estuaries are ephemeral features that are constantly being filled with sediment from both land and marine sources; and developed special sampling methods to measure big rivers and understand the movement and storage of sediment through these river systems throughout the world.

Perhaps the secret to Meade's success is his interaction with others. His interests range from art and theatre to the earth sciences, and his passion for observing the natural world flows over into the way he shares it with others. He has a talent for painting pictures with words and communicating scientific images, ideas, and concepts to the public.

Meade's career resembles a great river, linking, benefiting from, and influencing the lives and ideas of many distant and different scientists. He has mentored many young scientists and worked with scientists from the United States, Europe, Russia, China, Brazil, Canada, Venezuela, and elsewhere. He once remarked, "I was lucky, over the years, to work with exceptional colleagues. Looking back, perhaps the greatest joy was being able to interact with so many quick and fertile minds."

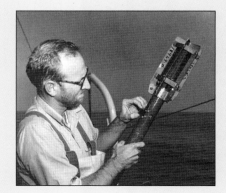

2000
The Great Lakes Science Center assembles more than 30 years of survey data on prey fish and fish populations to sustain the top predators in the Great Lakes.

2000
The USGS publishes the Global Seismic Hazard Map, a cooperative research effort among hundreds of scientists from all over the world.

2000
The USGS publishes the World Petroleum Assessment 2000, providing estimates of undiscovered oil and gas resources of all lands exclusive of the United States.

2000
The USGS begins a large-scale transformation of its traditional topographic mapping efforts. *The National Map* is a partnership-based effort led by the USGS to build America's infrastructure of geographic knowledge and ensure access to current, nationally consistent, complete, and accurate information.

Samuel N. Luoma

Many people shy away from complexities. USGS hydrologist Sam Luoma is not one of them. His work has had a profound impact because he challenges accepted paradigms and tackles the complexities of environmental processes. While Luoma has made substantial scientific contributions to a variety of areas, he has had particular influence with his studies on the effects of pollutants, especially metals, in aquatic environments. He

Sean Richards

has helped improve the design and function of national programs related to water quality in the United States and Canada. This work includes service as the first Lead Scientist for the CALFED Bay-Delta program, helping to provide the science to manage the San Francisco Bay and the Clark Fork River. His belief in the value of long-term datasets for ecological studies led him to play a major role in the conception, design, and implementation of the successful USGS National Water Quality Assessment Program.

Luoma gives his time and expertise generously, participating in many national and international committees, panels, and scientific societies. He is passionate about mentoring early-career scientists and has put special emphasis on enhancing diversity.

An exemplar of communicating science, Luoma has authored more than 180 peer-reviewed publications; written a textbook on environmental issues; and served as editor of professional journals and publication series. Under a Fulbright Fellowship, he is currently collaborating with colleagues at the Natural History Museum in London to write a book on environmental toxicology; this book is directed at helping managers, policymakers, and students to effectively link science and policy. Luoma's boundless energy, enthusiasm, creativity, insight, and eloquence in articulating the significance of science have made him an extraordinarily effective scientist and an outstanding public servant.

life and improved ability to forecast possible scenarios. The Hawaiian Islands face invasions of exotic species and decline of coral reefs. Other DOI bureaus are working together with the USGS to develop and coordinate long-term science strategies. The result will be the incorporation of DOI science priorities into USGS 5-year strategic plans and annual operational plans for major scientific programs and for all USGS disciplines—biology, geography, geology, and hydrology.

USGS and Native Americans—We Are All Related

The USGS cooperates with American Indian and Alaska Native governments to conduct impartial research or provide technical expertise on water and mineral resources; animals and plants of environmental, economic, or subsistence importance to Native Americans; natural hazards; and geologic resources. The USGS recognizes the need to learn from and share knowledge with Native peoples. USGS works with Tribes, Tribal organizations, Native professional societies, and the Bureau of Indian Affairs (BIA) to ascertain specific needs and develop responses that fall within the mission of the USGS. An American Indian/Alaska Native Coordinating Team has been established to develop policy and to coordinate USGS activities with Native peoples.

Within the context of the Department of the Interior's scientific priorities, the USGS provides information to the BIA to benefit American Indian and Alaska Native peoples and their lands. Other projects grow out of initiatives designed and conducted by USGS employees

in response to observed needs. USGS employees also assist American Indians and Alaska Natives by participating in organizations to promote interest in science among Native peoples and to help build and support communication networks.

The USGS encourages American Indians and Alaska Natives to pursue careers in science and seeks ways to facilitate the hiring of qualified students. For example, the Rosebud Sioux Tribe, Sinte Gleska University, and the USGS have been working together for several years to enhance science education for Native American students. Dur-

2001
September 11 terrorist attacks. USGS conducts studies of dust particles in the aftermath of the collapse of the World Trade Center.

David E. Click.

ing 2002, a group of Tribal and university professionals from water resources, land records, cultural resources, biology, and other fields gathered for 3 days of GIS training at Sinte Gleska University (SGU) on the Rosebud Sioux lands in Mission, S. Dak. James Rattling Leaf, from the SGU Sicangu Policy Institute, and USGS geographer Joseph Kerski conducted the workshop, which emphasized the use of digital spatial data, including computer maps, satellite imagery, and aerial photographs. Participants collected field data with Global Position-

ing Systems, brought field data into a GIS, analyzed base spatial data from theWeb, analyzed natural hazards and demography in the region and on the Tribal lands, and engaged in other activities. The goals were to bring different departments and organizations in the Tribal Government together to work on joint projects, and to help Tribal professionals make wise decisions about the Earth and its people.

Native self-sufficiency, economic development, and conservation are nurtured through Native decisions, and

these decisions can benefit from USGS data and analyses. The USGS recognizes the need to make its information available to Tribal Governments and to work with those governments and other institutions to advance data management capabilities. This information is available through printed reports, digital technologies, maps, workshops, and personal interactions with USGS employees. The USGS also recognizes that Tribal institutions have varying needs, interests, and capacities and strives to be sensitive to the unique circumstances of each of these institutions. Native peoples have unique understandings of nature, the landscape, and change in what is now the United States. The USGS wants to share its knowledge as well as to learn from Native perspectives that can further our scientific understanding.

Thomas C. Winter

Thirty years ago, little was known about how lakes and wetlands interact with ground water. Today, largely due to the research of Tom Winter, we have a better understanding of the different processes controlling lake and wetland water levels and how these change over time in response to natural variation between wetter and drier conditions.

Winter started with numerical simulation of the interaction of lakes and wetlands with ground water, designing mathematical models of physical processes to represent a variety of real-world settings, but fieldwork was essential to discover how closely the models matched reality. Collaborative research at several sites over the past decades has enabled Winter to demonstrate several important findings. These include the importance of evaporation in the dynamics of these waters, as well as the role of ground-water interaction with lakes and wetlands in determining water-level changes and the length of time needed for water and solutes to move into and out of these aquatic systems.

Winter's studies of evaporation and ground-water flow at the various study sites over many years have provided the foundation for a new conceptual framework for relating the hydrological characteristics of lakes and wetlands to the living things that inhabit them. Today, Winter's work not only serves as a guide to how hydrological process research about lakes and wetlands should be conducted but also provides lake and wetland managers with the scientific foundation for decisions they need to make about these aquatic systems.

2001
The Yellowstone Volcano Observatory opens.

2001
The NASA/USGS Extreme Storm Hazards Map is published.

2001
USGS research finds that contaminants play an important role in the decline of amphibians in California ponds and streams. Pesticides from the Central Valley are transported to the Sierra Nevada on prevailing summer winds and may be absorbed by frogs, suppressing an enzyme that is essential to their nervous system.

2001
The new, online National Water Information System (NWISWeb) allows users to access more than 100 years of real-time and historical USGS water information with the click of a mouse.

Alaska Volcano Observatory

Since 1760, more than 260 eruptions have been reported from the 41 historically active volcanoes in Alaska. In 1988, the USGS, the University of Alaska Fairbanks Geophysical Institute, and the Alaska Division of Geological and Geophysical Surveys came together to form the Alaska Volcano Observatory (AVO), a cooperative effort to minimize the impact of volcanic eruptions.

Two powerful tools help the AVO keep tabs on volcanoes: seismic monitoring networks and satellite data monitoring. By carefully watching for changes, these systems can provide warnings of potential eruptions, deliver important information during volcanic activity, and help build the growing scientific understanding of volcanoes and volcanic processes.

Recent explosive eruptions and periods of unrest at some of the active volcanoes have significantly affected air traffic over the North Pacific; Alaska's local communities; and oil, power, and

fishing industries. These impacts have prompted significant expansion of the observatory's volcano-monitoring capabilities. The number of volcanoes monitored by seismic networks has increased from 4 in 1995 to 27 in early 2004. The AVO has also increased its capabilities to acquire, store, and analyze data for tracking volcanic activity. With each new seismic network, volcano-hazard assessment, eruption, and period of volcano unrest, scientists increase their ability to provide the aviation industry and the general public with timely warnings and crucial information to deal with Alaska's active volcanoes.

Science Around the World

International work has been an integral part of the USGS for 122 years. The first international study by a USGS scientist was in 1882, when Clarence Dutton studied volcanoes in the Kingdom of Hawaii. Cooperative work on U.S./Canada border areas with the Geological Survey of Canada began 2 years later. USGS international activities assumed greater prominence in 1897, when a USGS geologist and a hydrographer were detailed to the Nicaraguan Canal Commission to study the proposed canal route between the Atlantic and Pacific Oceans, and in 1898, when USGS geologists were sent to investigate the mineral resources of Cuba and the Philippines.

In 1944, the first USGS employee arrived in the Kingdom of Saudi Arabia in response to a request from the King for a foreign technical expert, not affiliated with an oil company, to advise him on the Kingdom's natural resources (particularly water). Early work included some of the first geologic traverses of western and central Arabia, by truck, camel, donkey, and on foot. The USGS had a presence in the Kingdom until 2003, and scientific activities included studying the geology, water resources, mineral resources, and seismicity of the Arabian Peninsula. The USGS involvement evolved from planning, managing, and conducting the scientific work to acting as technical advisors to the newly created Saudi Geological Survey. Over the years, USGS employees helped to train and mentor many Saudis in a wide range of jobs, contributed significantly to the geologic knowledge base, identified numerous mineral discoveries, and helped in the creation of the Saudi Geological Survey.

Today, the USGS is involved in a wide range of international work that engages all of its scientific disciplines. These include efforts to control the spread of invasive species, provide global data-sets via the Landsat satellite system, cre-

NASA

2002

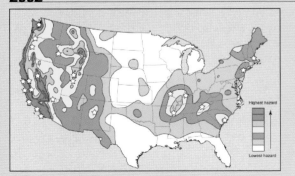

The Advanced National Seismic System (ANSS), and associated products like ShakeMap and the National Seismic Hazard Map, brings the capability to rapidly deliver useful information to emergency managers in real time. ANSS is installed in several cities—Seattle, Los Angeles, San Francisco, Salt Lake City, Memphis, and Anchorage—that are identified as high risk.

ate tools to manage ground and surface water resources, and develop assessments of global energy and mineral resources. Some recent international efforts are described below.

The USGS developed the Volcano Disaster Assistance Program (VDAP) after the deadly eruption of Colombia's Nevado del Ruiz in 1984, which killed 23,000 people. This program, a partnership between the USGS and the Office of Foreign Disaster Assistance of the U.S. Agency for International Development (USAID), has formed the world's only standing volcano-crisis response team, a group that is able to quickly mobilize to assess and monitor hazards at volcanoes threatening to erupt. The VDAP team works with local scientists and technicians to help them provide timely information and analysis to emergency managers and public officials. In addition to deploying crisis-response

teams, VDAP conducts training exercises and workshops in volcano-hazards response for foreign scientists and emergency management officials. VDAP has proven to be a highly effective international effort and, as an example, is credited with saving many lives and hundreds of millions of dollars worth of U.S. equipment by providing evacuation warnings during the 1991 eruption of Mount Pinatubo in the Philippines.

In 1998, flooding and landslides associated with Hurricane Mitch devastated large parts of Central America (Nicaragua, Honduras, El Salvador, and Guatemala). With the support of

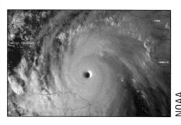

NOAA

USAID, the USGS responded with a coordinated effort that integrated its diverse capabilities. These included installation of telemetered streamgages to provide early flood warnings, hazard mapping to delineate areas at risk for landslides, geospatial coverages (topographic, aerial, and satellite) and installation of local GIS

Woods Hole Science Center

The underwater areas off our Nation's shores may seem like a different world, but our relationship with coastal and marine environments affects both the health of the land we live on and that of the oceans. Getting to know the scientific links among the land, coastal zone, and the oceans is the first step to a good relationship. Since 1962, USGS scientists at the Woods Hole Science Center have explored the coastal-marine environment, from the coasts (including beaches, estuaries, rivers, wetlands, and large lakes), to the shallower slopes that surround the continent and the deep ocean floor. Results of USGS studies are providing information so that Federal, State, and local agencies and the public can make informed decisions about resource management, hazards mitigation, human health, and environmental quality. This research addresses both natural geologic processes and human influences to improve our understanding of the Nation's coastal and marine regions, to provide a comprehensive scientific knowledge bank of that understanding, and to develop the ability to predict coastal and marine geologic processes.

Of the many accomplishments of USGS scientists at Woods Hole, one of the best known is the partnership with the Massachusetts Water Resources Authority (MWRA) and other agencies during the court-mandated cleanup of Boston Harbor, once declared the most polluted harbor in the Nation. In 1989, the USGS joined with MWRA to address more than 300 years of metropolitan waste disposal. Together, they have worked to understand and predict the fate of contaminants introduced to Massachusetts' coastal waters. USGS work helped MWRA locate its offshore sewage outfall discharge, documents the continuing improvement of the harbor environment, and monitors the harbor cleanup.

2002
The National Petroleum Reserve–Alaska (NPRA) oil and gas assessment is released, as is the oil and gas assessment of the basins in the Rocky Mountain region.

2002
Scientists at the USGS report that pharmaceuticals, hormones, and other organic waste-water-related chemicals have been detected at very low concentrations in streams across the Nation.

2003
Work pioneered by USGS in developing improved online data search techniques for scientific data is applied in Science.gov.

systems, and biological studies of the hurricane's impact on coral reefs, mangroves, and aquaculture.

In 1998, areas just to the east of Istanbul, Turkey, were devastated by powerful earthquakes that took the lives of more than 20,000 persons. The USGS responded with a variety of cooperative efforts with Turkish colleagues that included

- Geological evaluation of the structures that controlled the earthquakes and an assessment of the probability of future devastating earthquakes,

- Evaluation of specific buildings to determine how they responded to the earthquakes and their future viability, and

- Mapping of critical urban areas that showed anticipated damage resulting from subsequent earthquakes.

Because the geologic faults in Turkey are very similar to those in California, many aspects of this work could be directly applied to the evaluation of earthquake risks in the United States, a transfer that illustrates how USGS international efforts complement its domestic work.

A number of USGS international activities have involved a mixture of the scientific and the diplomatic. The best examples are provided by USGS efforts with regards to water and earthquakes in the Middle East. In a region long troubled by political differences, the USGS has led regional projects focused on developing common ground-water data sets that are essential for the management of shared ground-water resources. Similarly, the USGS has led efforts in the Middle East to foster the sharing of seismological data with the goal

Serkan Bozkurt

of developing better strategies to mitigate shared earthquake hazards. Over many years, both efforts have proven effective in fostering regional cooperation, elevating the region's technical capacity, and promoting interactions among the region's technical agencies.

In the future, international activities will increase in importance for the USGS. For example, scientific and technical information will be needed to support the U.S. commitment to increasing the availability of clean water to the developing world. As the world population continues to increase, there will be increased need for monitoring man's interaction with the land, an activity the USGS is currently doing in its support of the Famine Early Warning System. In addressing these and other international issues, the USGS will be able to build on its domestically developed core capabilities of integrated science involving the biological, geographic, hydrologic, and geologic disciplines.

2003

The USGS leads the Expanded Electronic Government initiative, a departmentwide GIS enterprise system to provide benefits and cost efficiencies in delivering geospatial data and analysis tools to internal and external customers.

2003

The Geospatial One-Stop portal is launched.

2004

The USGS provides access to publications through the Publications Warehouse.

2004

The USGS celebrates its 125th anniversary.

To mark the anniversary, this Circular highlights 125 years of science for America.

An Exciting Future

Linking Science and Society: The USGS Science Impact Program

USGS science is used every day by people throughout the United States and around the world to address a broad spectrum of environmental, natural-resource, and hazards-related issues. In recent years, technology, decisionmaking processes, and data accessibility have changed dramatically. Decisions are increasingly decentralized, analytical capabilities are expanding, and data accessibility is growing. Decisionmakers and citizens

Corbis Images

demand accessible, timely, and integrated science products that objectively address policy-relevant issues. As the world changes, USGS must increase and strengthen efforts to make its research and information as useful as possible. The Science Impact program is a focused effort to improve and expand the use of USGS science information to support decisionmaking at DOI; at other Federal, State, and local government organizations; and by the public. Science Impact will build on the good work that USGS is currently doing to link science with societal decisions and will focus its research efforts on designing and improving innovative methods and processes to increase the effective use of science information. Science Impact focuses on three principal activities:

- Science Synthesis encompasses identifying, developing, and evaluating needs and opportunities for science to support decision-making. Societal issues, disputes, and problems are linked with current and future science capabilities to determine the context in which science can support decisions, resulting in the identification of science questions and information needs.

- Tool and Product Development includes collaborating with public, academic, and private partners to develop integrated multidisciplinary tools, products, models, and processes that describe the physical, economic, and social implications of alternative decisions.

- Science Impact Education and Training relates to training activities designed to improve the interface among scientists, decisionmakers, and citizens and to facilitate informed and effective use of USGS science information, tools, models, and products.

Science Impact draws upon skills that extend beyond the traditional USGS disciplines. To gain expertise and capacity, the USGS is establishing Science Impact Partnerships with external organizations that have demonstrated capabilities in these areas. Five Science Impact Partnerships have been established initially:

- Science Impact Laboratory for Policy and Economics, University of New Mexico

- Science Impact Laboratory for Urban Systems, University of Pennsylvania

- Science Impact Center for

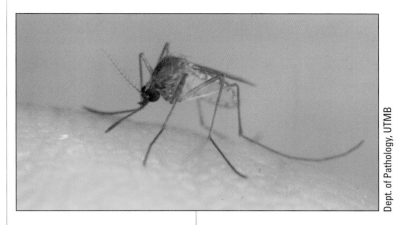

Dept. of Pathology, UTMB

Visualization and Delivery, Prescott College

- Indigenous Knowledge Center for Education and Science Impact, Sinte Gleska University

- MIT–USGS Science Impact Collaborative, Massachusetts Institute of Technology

Through innovative integrated natural and social science research, the development of university partnerships, and collaboration with scientists, decisionmakers, and other stakeholders, USGS scientists are maintaining and enhancing the commitment to its core values of science excellence, science leadership, and science impact to provide the information needed for informed decisions now and for the next 125 years.

New, Emerging, and Resurging Diseases

USGS is recognized as a leader in topics related to human health, such as toxic substances in water and in geologic materials, vector-borne diseases, and zoonotic diseases—animal diseases that can be transmitted to people. This research, conducted in collaboration with many of the Nation's leading health research organizations, including the National Institutes of Health and the Centers for Disease Control and Prevention, attempts to understand how exposure to chemical contaminants and pathogens in soils, water, and air affects the health of people throughout the world. For example, research by USGS scientists on a dev-

Employees of the Future

From today's perspective, the USGS employees of the future will need to be proficient in many different areas. With the knowledge, tools, and technologies of today's employees to build upon, the bar is set high for future employees. With so much to know, the ability to work as a team on multifaceted projects is a must.

Future USGS scientists will improve understanding about human health issues, work on the increasingly urgent issues of water quality and quantity, provide foundational research on invasive species, determine disease vectors and potential impact, continue to focus on hazards mitigation, provide a greater understanding of complex Earth systems and Earth system dynamics, research nontraditional sources of energy, and address the impact, cost, and environmental impact of alternative fuel sources. They will use long-term data to develop predictive models. Because they will understand the political, societal, and even global impacts of their science, they will translate their work for a wide variety of audiences, including other scientists, students, teachers, government agencies, decisionmakers, and everyday people around the world.

The way they conduct their work will be revolutionized, crossing borders and boundaries among agencies, organizations, countries, and continents in a virtual environment that promotes collaboration, partnership, and knowledge sharing as never before. Wherever they are, they will have access to their work and all the information they could ask for on their area of research. Business and communications skills will be second nature to ensure the success of current and future science. The employees of the future will come from a wide variety of cultures and perspectives, fostering creative solutions for both old and new problems. Because they will utilize integrated approaches and develop relationships in other offices, agencies, and political and social realms, they will find solutions more quickly and deliver information to millions of people in real time.

The USGS has been providing science for a changing world for 125 years, matching the abilities and knowledge of its employees to the progress of science and technology. The dedication and skill that make the USGS today a valued source of information for sound decisions will continue as we enter our next 125 years. The future of the global community provides unprecedented opportunities for the people of the USGS to contribute credible, impartial, excellent science for the betterment of the Nation and the world.

astating degenerative kidney disease and associated cancers affecting thousands of people in the Balkans has shown that the problem may be caused by drinking well water that has been in contact with lignite, a form of coal. Applying this observation to the United States has revealed a possible link between well water in contact with lignite and a high incidence of certain cancers.

There is also increasing concern about the emergence of new infectious diseases afflicting humans and the resurgence of previously conquered diseases. Within the United States, the assault on humans includes invasion by exotic diseases such as AIDS, West Nile virus, and monkeypox; the appearance of novel diseases from indigenous sources such as Legionnaires disease and hantavirus infections; the resurgence of diseases once thought be "conquered," such as tuberculosis and malaria; and geographic changes in the distribution of historic diseases such as sylvatic plague and tularemia.

It is increasingly clear that studying the human aspects of disease is only one part of the solution—we must also understand how wildlife populations can serve as reservoirs for disease and how insect vectors affect disease transmission

Wyoming Game and Fish Department

(such as the deer tick responsible for transmitting Lyme disease to humans). Many of these diseases also afflict wildlife and domestic animals. The USGS is charged with addressing health issues and diseases of free-ranging wildlife across the Nation. The 1999 emergence of West Nile virus in the Western Hemisphere marked the first time that public, agriculture, and wildlife health scientists joined forces to tackle and control a common enemy. USGS scientists quickly collaborated with public health agencies to track this mosquito-borne virus of birds as it began spreading across North America killing many wild birds, horses, and humans.

USGS investigations have detected a number of novel wildlife diseases that have resulted in management actions to prevent human exposures when zoonoses such as tularemia, chlamydiosis, and erysipelas have been found. As new disease threats such as West Nile virus, chronic wasting disease, and monkeypox

emerge, and old ones such as avian influenza, foot-and-mouth disease, and plague resurface, the collaborative work of many agencies is needed to grapple with these complex diseases. Timely and accurate diagnosis of wildlife illness and mortality is critical to achieving effective disease control and prevention. USGS will continue to expand our understanding of wildlife disease dynamics and risk analysis through research, surveillance, and effective sampling. By increasing our understanding of the complex interactions among wildlife, humans, geologic materials, and ecosystems, the USGS will continue to provide information and analysis to keep people and ecosystems healthy and thriving into the next century.

Tracking Emerging Contaminants

Chemicals used every day in homes, industry, and agriculture can enter the environment through our wastewater and other mechanisms. These chemicals include human and veterinary pharmaceuticals (for example, antibiotics and synthetic hormones), detergents, disinfectants, fragrances, lotions, plasticizers, fire retar-

dants, insecticides, and anti-oxidants. The presence of these chemicals raises several issues for which science can provide answers: These chemicals are not routinely monitored in our water resources; many have known or suspected ecological or human health effects; and drinking water standards have not been developed for them. The USGS is developing new laboratory methods to measure minute quantities of these chemicals; defining the environmental occurrence of these compounds; developing knowledge of the biological and chemical processes that affect contaminant movement, transformation, and persistence in the environment; and assessing the effects these contaminants have on organisms and biological communities within aquatic ecosystems. For example, a group of commonly used flame retardants appears to be rapidly increasing in concentration in the environment. Recent evidence from European and U.S. investigators has shown that concentrations of these chemicals are doubling in human breast milk at the rate of every 2–5 years. Information about the environmental fate and effects of these compounds is critically needed so that scientists can determine at what

Diversity

The USGS owes 125 years of success to the creative discoveries of its people. As we head into the next 125 years, a diverse workforce will play an ever-increasing role in maintaining an atmosphere of creativity and bringing new discoveries to the world.

Increased numbers of women, minorities, people with disabilities, and people from a wide range of backgrounds and beliefs will result in greater learning, enhanced creativity, improved products, and increased responsibility to the customers that the USGS serves. As technology continues to erase borders, a variety of cultures and perspectives will be increasingly important to daily life at the USGS. Research will often be a global effort, providing people with greater resources and a broader understanding of each other and our planet. USGS employees will not only have the ability to inform and cooperate with many different types of people, they will also represent and reflect the demographic variety of the Nation.

The Nation's changing demographics require that the USGS continually evolve to meet the country's needs. The USGS must develop science products that are responsive to the needs of constituents and that are relevant to users. Science requires a global perspective and an appreciation and understanding of many cultures.The snakehead fish, for example, is part of the diet in some cultures but is an invasive species in the United States.

Over the last 10 years, progress toward a more diverse USGS has been made. The number of women advancing at the USGS is on the rise, but there is still work to be done. Minorities and people with disabilities still lag behind in terms of representation and advancement in the workforce. The USGS will continue to show commitment and investment in the future through participating in educational partnerships, fostering mentoring programs, and providing role models to students.

EROS Data Center

The EROS Data Center (EDC) is a world leader in the exploration of our changing planet. In this 600-person office in Sioux Falls, S. Dak., remotely sensed images of the Earth are collected, stored, and analyzed. The work done at EDC enhances our understanding of the Earth, how it changes over time, and the implications of those changes for people and ecosystems worldwide. EDC maintains one of the largest collections of remotely sensed images of the Earth's land surface, including images from Landsat, NOAA weather satellites, NASA satellites, and USGS aerial mapping photography. The aerial and satellite data archive now contains more than 38.5 million images and is growing by close to 10 million images each year. EDC staff members control satellites, distribute land information products, and develop techniques for providing scientists, policymakers, and educators around the world with access to the ever-growing collections of data. Representatives from almost every science discipline work with foresters, satellite engineers, and data processing specialists to handle the data, analyze features, and develop new systems for observing changes in the Earth. They study changes such as urban sprawl and respond to disasters

such as wildfires, floods, tornadoes, and hurricanes. They sometimes work globally, for instance studying the loss of wetland in Southern Iraq or the deforestation of tropical rainforest in Bolivia. As EDC observes our ever-changing planet, it looks to the future, encouraging and educating students through a variety of student mentoring and classroom demonstrations, higher education consortiums, collaborative research projects, and visiting scientists and internships.

point the compounds will reach concentrations that are toxic to humans and biotic resources.

For all of these potentially toxic substances, USGS scientists are working to develop the scientific basis for assessment, restoration, and monitoring of habitats that are contaminated by mining, agriculture, forestry, wastewater, industry, and chemical control agents. Scientists are using laboratory test to investigate substances that are lethal to plants and animals or that suppress immune function, disrupt the endocrine system,

deform embryos, or impair reproduction or neurological functioning. USGS research is providing methods and guidance for future monitoring and assessment of these types of environmental contaminants and establishing the needed foundation for setting priorities for further study of sources, pathways and effects. The data and information provided by USGS are valued by both scientists and policymakers working to keep our environment healthy for all.

Water Availability

Water availability is a topic of increasing concern in virtually all parts of the Nation for a variety of reasons, including recent and ongoing droughts and the competition for water among jurisdictions and among ecological, urban, agricultural, and industrial uses. A number of recent changes exacerbate the problem. Population growth nationally, and particularly in the more arid Southwest and coastal parts of the Nation, presents major challenges for State, regional, and local authorities who must find reliable, high-quality water supplies to deliver to their communities. Public values regarding the environment, coupled with the Endangered Species Act, have led to the need for reallocation of water between traditional off-stream uses (irrigation, thermo-electric power, public supply, and industry) and instream needs of the biological community. Ecosystem restoration is a growing national interest, and, in many cases, reallocation and changes in the storage of water and timing of releases are central to ecosystem restoration plans.

In addition, although agricultural water use is fairly stable when averaged across the Nation, new demands for water

for irrigation, animal feeding operations, and aquaculture are having a significant impact on the water budget in many humid areas such as the Mississippi Delta and Atlantic Coastal Plain, putting stress on regional water supplies. The pumping of ground water is an increasing factor in the Nation's water supply, and its use is leading to impacts that may result in the need to slow the growth or even decrease the rate of use of ground water in some areas. The impacts of increased ground-water development can include increased costs due to the need to drill new wells, intrusion of saline water (from the ocean or from deeper units), depletion of base flow in streams, increased summer temperatures in small streams due to decreased ground-water inflows, and land subsidence. In areas including parts of the High Plains, Mississippi Delta, Chicago/Milwaukee area, Southwestern basins (such as Las Vegas, Middle Rio Grande, Tucson, Phoenix, Mojave Desert), and parts of the Atlantic Coastal Plain, the rates of pumping are not sustainable. New solutions must be sought in terms of either new supplies or decreases in use.

Finally, there are beginning to be clear indications that climate warming is leading to changes in the type of precipitation in some parts of the Nation, generally less snow and more rain, and to earlier melt of winter snows. As a result, the amount of natural storage of water in snowpacks is decreasing. These changes, along with other potential shifts in hydroclimatic conditions, could result in fundamental changes in the overall amount of the resource that is readily available for various offstream and ecological uses.

USGS science will be key to resolving these issues, providing the science needed to ensure that water is available to support human and ecological communities in the years ahead. Good planning in the public or private sector depends on good information that moves knowledge forward but always stays mindful of the remaining uncertainties. Disagreement and misinformation about water have created decision gridlocks over water and water-related ecosystem restoration projects that can harm both the environment and the economy. The USGS is ideally poised to help the Nation break these gridlocks through improved understanding and communication of the principles and facts that all the parties need.

Energy and Minerals for the Future

All nations continually face decisions involving the supply and utilization of raw materials, substitution of one resource for another, competing land and resource uses, and the environmental consequences of resource development. Global use of mineral and energy resources will continue to increase for the foreseeable future because of the continuing increase in global population and the efforts to improve living standards worldwide. In today's global economy, a nation's economic security depends on access to adequate energy and mineral resource supplies from a variety of domestic and international sources. Today, more than 70 chemical elements and dozens of minerals are mined and produced from more than 100 different deposit types and geologic environments. USGS researchers are currently focused on assessing global undiscovered resources of copper (important in the electronics industry), platinum-group metals (critical as catalysts in the automobile, chemical, and petroleum industries), and potash (one of three indispensable fertilizer minerals required for food production).

Geologists in the USGS have been working for many years to understand how different ore deposits were formed, with the goal of aiding in the recognition of undiscovered resources, in the efficient extraction of the resource, and in the ability to assess undiscovered resources. The work of identifying mineral deposit models has been going on for many years, and now more than 100 models have been defined. Research continues to develop new deposit models, to clarify existing models, and to refine our ability to conduct mineral resource assessments that are accurate and cost effective.

Assessments of world and national energy resources are changing as the needs and the information available have changed and as new types of energy resources are recognized. Over the past few years, basic geologic understanding of coal-bed methane occurrence, characterization, and formation has provided the information needed to help make this once-unconventional resource a viable contributor to today's energy mix; coal-bed methane is now 8 percent of all natural gas production in the United States.

Although conventional accumulations of oil and gas

Mobile Computing

Operating systems that run on compact hand-held devices such as Personal Digital Assistants and "Palmtop" or "Pocket" personal computers (PCs) have the potential to replace paper forms in many USGS field data collection activities. Currently, many USGS offices use such devices to communicate with data loggers and data collection platforms (DCPs) and to log field data on digital (virtual) forms. Software programs used on these mini-PCs enable communication with a variety of field instruments that collect surface-water, ground-water, water-quality, and biologic data. The benefits of this new technology include improved productivity and data quality.

resources remain very large and work identifying these continues, increased attention is being focused on unconventional resources such as gas hydrates. USGS is partnering with a number of international consortia to study gas hydrates, a potentially huge energy resource. The estimated amount of natural gas in the gas hydrate accumulations of the world greatly exceeds the volume of all known conventional gas resources.

While gas hydrates hold great potential as an environmentally friendly fuel for the 21st century, the geologic challenges of understanding their formation and distribution and the technical challenges of

realizing them as a resource are substantial. Additional research is required to understand and develop new techniques to quantify their distribution in nature. The USGS has developed a state-of-the-art geotechnical laboratory capability to determine the pressure and temperature conditions necessary to induce gas hydrate formation and test hydrate samples recovered from Arctic and offshore coring operations. In December 2003, an international consortium including the USGS proved that gas hydrates are technically recoverable, though economic recoverability has yet to be determined. Other studies underway will help determine the amount of technically recoverable gas hydrate present in the United States and worldwide. Through innovative research and modeling efforts like these, USGS scientists will continue to provide essential information on mineral and energy resources for a healthy economy and enhanced quality of life.

Making Wise Use of New Technologies

USGS scientists are taking advantage of new technologies to revolutionize the methods

used to collect field data and to provide greater accessibility to our data and information.

New Water-Quality Sensors

To provide real-time answers for water-quality management decisions, including early-warning monitoring for intentional or accidental releases of hazardous materials in water, the USGS is working with partners in the research community and local governments to add new, advanced-technology water-quality sensors to our existing real-time data collection network. Submersible analyzers capable of measuring plant nutrients—nitrate and phosphate, for example—by standard methods became commercially available to marine scientists in the early 1990s. Over the past 5 years, USGS scientists have begun to use such analyzers for nutrient analysis in remote locations. Sampling rates as often as every 15 minutes over periods ranging from a few days to several weeks provide information about changes in nutrient concentrations that would be difficult or prohibitively expensive to obtain by conventional means. In addition, real-time

chemical analysis eliminates costs and potential errors associated with sample processing, shipping, and storage.

The USGS has also leveraged new technology to collect field data and to make scientific data and reports easily accessible to the public. The USGS National Water Quality Assessment Program (NAWQA) *<http://water.usgs.gov/nawqa/data>* has produced some of the most widely used water-quality data in the world and has consistently been Google's top pick for water-quality data over the last several years. The public can retrieve subsets of these data and view or export them in popular formats such as Excel and commercial-off-the-shelf (COTS) software from the Oracle Corporation. Using COTS software simplifies and lowers the cost of long-term management of the system. Users can make interactive maps of water-quality concentrations for hundreds of chemi-

cals in wells or streams, as well as graph the concentration range for hundreds of chemicals across dozens of basins in the country. With this Web-based interactive system, more than 10,000 graphs can be specified and created. New data are aggregated frequently and automatically from about 50 servers across the country by using integration COTS software from Informatica Corporation.

The Future of Streamgaging

Streams and rivers provide water for important uses including drinking water, navigation, and recreation. Measurement of river flow is needed to determine the amount of water available for these different uses and to monitor high flows and predict the frequency of flooding. The basic method for measuring streamflow in the United States has been virtually

unchanged for over a century, but this method is both expensive and, during large floods, potentially unreliable, unsafe, or impossible. Recognizing these problems, the USGS is investigating new and emerging technologies that may become the basis for a more cost-effective, accurate, safe, and robust method of streamgaging.

The most promising technology that can measure streamflows without having to touch the water is radar. USGS has demonstrated that different kinds of radars can measure the channel cross-sectional area and the surface velocity of rivers. Low-frequency radar antennas, identical to ground-penetrating radar (GPR), can be suspended above the river from a bridge or cableway, transmit a signal through the air, water, and bed material, and receive the return signals with sufficient clarity to define the channel cross section. Numerous experiments have shown that in the absence of highly conductive water (high conductivity destroys the strength of a radar signal in water), GPR can measure depths to within a few centimeters and cross-sectional area within a few percent.

An ideal system would be mounted on the river bank and scan the river bed continuously, but this presents significant challenges to radar signal interpretation and processing, especially the location from which return signals are received. The USGS has used a bank-operated light cableway extended across the river on which the GPR antenna is suspended and has mounted the antenna on the underside of a helicopter and flown across the river cross section. Surface velocity of rivers has been measured by using a pulsed Doppler radar system developed by the Applied Physics Lab at the University of Washington. These techniques measure the velocity of the river current at or very close to the surface of the water and can scan the width of the river from a single point on the bank. Surface velocity can then be converted to mean velocity, combined with the GPR-mea-

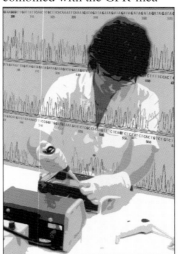

sured cross-sectional area, and used to calculate the stream discharge, all without having to touch the water. The results of experiments using these methods have been encouraging. The USGS will continue to explore new ways of measuring our Nation's streams safely and reliably.

Genetics and Molecular Biology

Recovering species that have been reduced to low numbers and small, scattered populations is a daunting and complex task for Federal and State natural resource managers. USGS scientists use genetics and molecular biological tools and analyses as increasingly routine means to assist managers in reducing population declines and bringing animal and plant species back from the brink. Because virtually every organism has unique sequences of DNA (deoxyribonucleic acid), genetic markers can be used to identify individuals and assign them to families, populations, and even in some cases their place of birth. Genetic markers can be used as a nonharmful way to track individual animals in the environment through time, or to trace the source of harmful bacteria in ground water.

Genetics research is on a fast track, and new uses of genetics and molecular biology promise additional important applications. Genomics—the study of the entire genetic makeup of a species—is now being proposed as a means for determining the functions (or expression) of individual genes or genes acting in concert to produce enzymes and other protein products. Scientists seek with new genomics technologies to understand the basis for disease or disease resistance, how pollutants damage DNA, or under what environmental conditions genes are turned "on" or "off" in their functioning. USGS scientists also propose genetic means for applying tools of microbial biotransformation and bioremediation to explore means to use microbes in habitat restoration and recovery. The techniques of genetics biotechnology work on anything containing DNA or RNA (ribonucleic acid). Unique genetic markers are discovered weekly and are added to international computer databases for sharing with researchers worldwide.

USGS genetics research is undertaken at the request of primary clients in DOI—the Fish and Wildlife Service, the National Park Service, the Minerals Management Service,

the Bureau of Reclamation, and the Bureau of Land Management—and other State and Federal agencies. The questions that managers pose often lead scientists down intriguing paths, such as whether the extinct blue pike was a separate species (our data indicate it was not), whether two invasive carp species in the Midwest can successfully mate and hybridize (they can), or whether black bear population numbers can be estimated from hair samples they leave on barbed wire (yes, and the bears can be tracked in the wild by this means as well). As exciting and satisfying as these genetics discoveries are, USGS scientists are just beginning to reveal the potential for the use of genetics and molecular biology in conservation. As technology develops, scientists can only guess at the ways they can apply their research to solving natural resource problems.

Earthquake Predictability

In the 25 years since the USGS Earthquake Hazard Program was authorized by Congress, important advances have been made in the ability to identify regions where earthquakes are likely to occur and to fore-

cast the hazards associated with damaging earthquakes. Many of these advances have their roots in laboratory and theoretical developments at the USGS, which, along with targeted seismic and geodetic monitoring, have focused on understanding the basic physical processes associated with earthquakes. This research provides much of the scientific underpinnings of the USGS's important and widely used products, including national and regional seismic hazard maps, regional earthquake probability reports, short-term foreshock and aftershock forecasts, and building code recommendations. The USGS is at a critical juncture in the field of earthquake studies. New dense arrays of seismic and geodetic instrumentation are being deployed across the United States and around the world, including the Advanced National Seismic System, the Global Seismographic Network,

and several elements of the National Science Foundation's EarthScope project. In addition, satellite-based remote-sensing systems are providing exciting new tools for conducting global-scale observations of surface deformation, electromagnetic fields, and other critical geophysical observations. Paleoseismological work on key faults, aided by new high-precision dating techniques, is producing detailed earthquake histories that extend back thousands of years. This quantum leap in the quality and variety of geological and geophysical data can be used to improve the precision of earthquake hazards assessments and to broaden the suite of earthquake information products available to emergency managers.

Remarkably, this flood of new data also provides the opportunity to test the predictability of earthquake occurrence. Despite earlier optimism,

it remains unclear whether useful intermediate- and short-term earthquake predictions are feasible. However, reputable research teams are taking advantage of improved data sets and are developing schemes for predicting earthquakes based on, for example, patterns of small earthquakes, patterns of surface strain, and electromagnetic field variations. The number, scope, and boldness of these efforts are accelerating. The USGS has mandated responsibility to provide scientific and public leadership in this rapidly shifting field, evaluating earthquake prediction methods, providing guidance to the government on appropriate public policy, and managing public expectations when earthquake predictions arise.

While working toward the elusive goal of reliable, specific earthquake prediction, the USGS is developing a number of products that will further improve public safety. Examples include daily earthquake hazard maps showing the chance of earthquake shaking in the next 24 hours, which rely on the quantifiable ability of one earthquake to trigger another, and earthquake early warning systems that would use ANSS data to determine that an earthquake is underway and

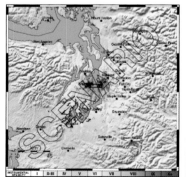

transmit that information before the shaking arrives at more distant sites.

Remote Sensing

The USGS has a proud history of acquiring, archiving, distributing, and using remotely sensed data. From aerial photography flown 50 years ago to more than 30 years of Landsat coverage, the USGS has established an important technology that is valued by users. In the future, land remote sensing will require new data sources, advanced approaches to analysis, and imaginative cooperation among Federal, international, and commercial data providers that will supply a massive range of information in real time about climate, hazards, ecosystems, national security—even traffic conditions. The USGS

will be a major contributor in ensuring the scientific integrity of remotely sensed data for analysis. Further, applications of remote-sensing technologies are advancing at an ever-increasing rate as new sensor platforms come online, user awareness of the systems' capabilities increases, and the costs associated with processing capacity continue to decrease.

New data sources and systems are certain to evolve in the near future. The science community both in government and in the private sector has come to depend on continuous, reliable data from key satellite data platforms, including Landsat, GOES, Terra, and Aqua; virtually all of them will need replacement within the next few years. A Landsat follow-on mission, an interferometric synthetic aperture radar mission, LIDAR, and

hyperspectral data sensors will provide high- to moderate-resolution data in broad spectral ranges, offering greater insight on conditions across the planet. The USGS looks forward to a leadership role in defining and making available data from those systems.

Not all of the new data sources will be Federal civilian programs. Partnerships with the commercial sector and international organizations will be important to remote sensing in the future. For example, the USGS commercial remote-sensing data contract program establishes a venue to consolidate access, processing, and distribution of remotely sensed data from a myriad of sensors and systems available from government, industry, and international sources. The increased cooperation, development of new techniques, and expanded

Patents and Inventions

Scientists sometimes find that they need a product that isn't on the market, or they suspect that a process nowhere used or taught will provide the results they need. In other cases, their interests and concerns, rather than their needs, lead them toward a new process or product. If the idea turns out to be novel, non-obvious, and useful enough, it may result in a patent.

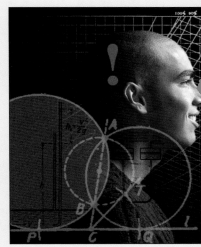

USGS has nearly 50 current or pending patents, including Infrasound Hazard-Warning Device for Night-Migrating Birds, Deep Aquifer Remediation System, and Mud Walking Shoe. Other examples are devices to control aquatic invasive species or remove toxic metals from water, and the hovercraft, for drilling core samples in wetlands and other environmentally sensitive regions.

The technology transfer program at the USGS *<www.usgs. gov/tech-transfer/>* combines the research capabilities of USGS scientists with the commercial development potential of the private sector. It encourages the adoption and use of USGS research products through partnerships. Technology transfer tools such as Cooperative Research and Development Agreements (CRADA) and patent licenses provide incentives for commercialization and use of technologies developed by USGS scientists. Patents, inventions, and CRADAs are a valuable outlet for the creative intelligence of USGS employees, benefit the public by creating jobs, produce financial benefits for the agency and its inventors, and offer one more means for USGS to enhance its public service.

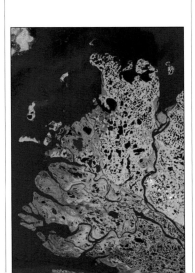

technical capabilities of satellite sensors will provide important tools for understanding a changing world.

Mapping the Nation in the 21st Century: *The National Map*

In the 1990s, sales of the USGS flagship product—the standard 1:24,000-scale, 7.5-minute topographic quadrangle map—steadily decreased, due largely to increased demand for cartographic data in digital form and diminished currency of the printed maps. By the year 2000, resource limitations and rising costs of traditional production methods resulted in an average age of 23 years for these 55,000 maps and an annual revision rate of less than 2%. In response to these conditions, the USGS, in 2000, initiated a large-scale

transformation of its traditional topographic mapping efforts. This reinvigorated program, *The National Map* <***www.nationalmap.gov***>, calls for partners throughout the Nation to work together in collecting geospatial data and in ensuring broad, open access to these data. It is a multi-sector effort to build America's infrastructure of geographic knowledge, bringing together State and local governments, the private sector, universities and libraries, and the public with Federal agencies led by the USGS to ensure access to current, nationally consistent, complete, and accurate data, tools, and applications.

The National Map takes advantage of emerging and coalescing technologies, including the Web and the widespread availability of geographic information system tools, to bring together base geographic

information from many sources to create integrated national data, while still providing access to partner data that provide higher resolution, more detailed coverage for limited areas and selected themes. The reliable, readily accessible, digital geographic framework of *The National Map* will foster place-based analyses of diverse types of information, assist in monitoring changes and detecting trends, and advance the discovery of relationships between otherwise seemingly independent phenomena and processes. When fully developed, *The National Map* will provide capabilities to view geographic information, access partner data, bring together data from diverse sources, develop customized maps that show current landscape conditions (the continuation of the USGS topographic map tradition), and

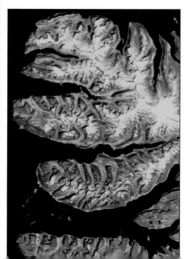

apply the tools of geographic analysis to understand the world in which we live and across which we interact.

Geographic Research for Tomorrow

Mapping, the foundation for place-based science and geographic analysis, continues as a vibrant and essential USGS activity, as demonstrated by the technical prowess and partnership strategy of *The National Map*. However, in the last 20 years the science discipline of geography has moved beyond describing the Earth with traditional topographic maps and has developed methods and tools of utility and sophistication that are widely used by the other natural and social sciences. A powerful combination of modern technologies related to place—satellite imagery, digital aerial photography, GIS, and satellite navigation (GPS)—enables computer assisted geographic analysis to create innovative possibilities for understanding the science of our earth and addressing complex environmental issues.

In the future, USGS geographic research will contribute new knowledge about the relation of nature and society

in such critical areas as human population growth, public health, environmental systems, water quality and availability, natural hazards (landslides, floods, earthquakes, coastal erosion), and climate change. USGS geography will also expand public involvement in decisionmaking by developing decision support systems (a type of computer model) to promote community-based environmental management. Decision support systems can deliver focused data and information on complex issues of natural science, often in the context of specific political, social, and economic structures, to local community leaders, land managers, and citizens. For both USGS clients and USGS scientists, the geographic perspective of natural science, the "where" factor, will become even more vital in integrating disparate data from a variety of sources because this perspective provides the common, easily visualized language of place.

Serving USGS Information to the Nation

The mission of the USGS is to conduct science and to make available the resulting knowledge and information. The USGS has a long history of providing high-quality service through dissemination of scientific information to its customers. Traditionally, this dissemination has been scattered throughout the organization, often making it very difficult for customers to locate the data and information that they need. The USGS will continue to use current dissemination mechanisms such as face-to-face visits, phone, fax, e-mail, and the Web to help us put information in the hands of the customer. However, new tools and approaches are needed. The USGS is undertaking a new approach to providing more efficient access to our data, information, and knowledge. This approach will depend on automated tools, a knowledge base of USGS science, and a core group of information professionals to bring the full breadth of USGS data and information to users

when, where, and how they need it. The goal is an integrated network of physical and virtual information offices that will enhance our ability to make our data, information, and knowledge available to anyone, anywhere at any time. By working toward this desired future state, the USGS will continue to serve as the Nation's primary integrated natural science and information agency.

For example, scientific publishing has been a key aspect of USGS science since its earliest days, but locating a specific USGS report can be difficult. To allow a single source for searching, viewing, and ordering USGS publications, in January 2004, the Publications Warehouse Web site was launched. Building on an internal database begun in mid-2000, the USGS Publications Warehouse *<pubs.usgs.gov>* provides complete and accurate bibliographic citations for 31 historic and current USGS series. The searchable database of reports and thematic maps currently contains more than 63,000 bibliographic citations, including numbered series begun as early as 1882. Citations and online documents are added regularly. Content ranges from more than 23,000 full text links to citations only. A number of Biological Resources dis-

cipline report citations, primarily historical research reports, are included. These citations include reports released by the Fish and Wildlife Service and the National Biological Service prior to the creation of the Biological Resources Division of the USGS in 1995. Plans are underway to add geospatial search tools, provide links to all online USGS series, and include citations and links to all publications with USGS authors. Eventually full text for all USGS reports will be available through the Publications Warehouse site. A single, comprehensive site facilitates access, assures reliability through multiple servers, and provides certification of USGS report content by serving from a single official, secure site.

Conclusion

Over the past 125 years, the USGS has evolved from gathering data with picks and pack mules to the use of remote sensors with real-time data access. Through a wealth of long-term data and research, the USGS has served the needs of society, the Earth, and its environment. It has responded to the changing needs of the Nation, expanding its traditional role of

geography, geology, and hydrology to include the assessment and monitoring of our Nation's biological resources. Today, the USGS brings these capabilities together to address complex issues in a comprehensive way that serves a myriad of customers and partners.

Technological capabilities have improved significantly during the past 125 years. For example, the ability to measure, monitor, and model the processes that lead to natural disasters has increased dramatically, as has the ability to communicate information about hazards to those whose lives may be affected. These innovations in tracking and communicating the changes in our dynamic planet, supported by a deeper understanding of Earth processes, enable the USGS to expand its predictive capabilities and point the way to a safer future.

There is a growing recognition of the connection between the environment and human health. The interface between wildlife and human health, the emergence of new contaminants in the environment, and the introduction and spread of invasive species all have significant and far-reaching implications for public health. In the years ahead, the USGS will play an increasingly important role in understanding

and ultimately solving these health issues.

When the USGS was established in the late 19th century, the Nation's natural resources of water, minerals, energy, and living things appeared limitless. The extinction of the passenger pigeon, drastic reductions in some animal and plant populations, energy shortages, and new concerns about the availability of clean water have brought a broader realization that resources are finite. However, human ingenuity, unlike other natural resources, is limitless, and USGS scientists are focusing their creative intelligence on providing the science needed to live fully yet lightly on the land.

After 125 years, the USGS continues to make a difference in the quality of life in America and abroad. The taxpayers' investment in basic science and monitoring at the USGS has paid a handsome dividend in science that everyone can use. Proud to be part of an organization with a distinguished history of unbiased science, the people of the USGS dedicate themselves to continuing their studies of the landscape, natural resources, and natural hazards. They look forward with fresh commitment to providing science for a changing world.

For More Information

Alley, W.M., Reilly, T.E., and Franke, O.L., 1999, Sustainability of ground-water resources: U.S. Geological Survey Circular 1186, 79 p.

Bohlen, S.R., Halley, R.B., Hickman, S.H., Johnson, S.Y., Lowenstern, J.B., Muhs, D.R., Plumlee, G.S., Thompson, G.A., Trauger, D.L., and Zoback, M.L., 1998, Geology for a changing world; a science strategy for the Geologic Division of the U.S. Geological Survey, 2000–2010: U.S. Geological Survey Circular 1172, 59 p.

Cowardin, L.M., Carter, V., Golet, F.C., and LaRoe, E.T., 1979, Classification of wetlands and deepwater habitats of the United States: U.S. Department of the Interior, Fish and Wildlife Service, Washington, D.C., 131 p.

Frazier, A.H., and Heckler, Wilbur, 1972, Embudo, New Mexico, birthplace of systematic stream gaging: U.S. Geological Survey Professional Paper 778, 23 p.

LaRoe, E.T., Farris, G.S., Puckett, C.E., Doran, P.D., and Mac, M. J., eds., 1995, Our living resources; a report to the Nation on the distribution, abundance, and health of U.S. plants, animals, and ecosystems: Washington, D.C., National Biological Service, 530 p.

Mac, M.J., Opler, P.A., Puckett Hacker, C.E., and Doran, P.D., eds., 1998, Status and trends of the Nation's biological resources: Reston, Va., U.S. Geological Survey, 2 vols., 964 p.

Rabbitt, M.C., 1989, The United States Geological Survey—1879–1989: U.S. Geological Survey Circular 1050, 52 p.

Robinson, G.D., and Spieker, A.M., 1978, Nature to be commanded...Earth-science maps applied to land and water management: U.S. Geological Survey Professional Paper 950, 95 p.

Thompson, M.M., 1981, Maps for America [2d ed.]: Reston, Va., U.S. Geological Survey, 265 p.

U.S. Geological Survey, 1904, The United States Geological Survey—Its origin, development, organization, and operations: U.S. Geological Survey Bulletin 227, 121 p.

Vigil, J.F., Pike, R.J., and Howell, D.G., 2000, A tapestry of time and terrain: U.S. Geological Survey Geologic Investigations Series Map I–2720, scale 1:3,500,000, 24-p. text.

Winter, T.C., Harvey, J.W., Franke, O.L., and Alley, W.M., 1998, Ground water and surface water; a single resource: U.S. Geological Survey Circular 1139, 79 p.

<www.usgs.gov>

<biology.usgs.gov>

<geography.usgs.gov>

<geology.usgs.gov>

<water.usgs.gov>

Figure Information

Front cover. Artist: Maura Harrison, USGS.

Back cover. Artist: Vitmary Rodriguez, USGS.

p. 1. The Organic Act establishing the U.S. Geological Survey. Photograph courtesy of Library of Congress.

p. 2. City, river, and drilling rigs. Artist: John M. Evans, USGS.

p. 3. Leadville, Colo., mining district, subject of an early mining geology study, 1879. USGS.

p. 4. Alabama shad. Artist: Duane Raver. Illustration courtesy of U.S. Fish and Wildlife Service.

p. 4. Major John Wesley Powell at his desk in the Adams Building on F Street, NW., Washington, D.C., ca. 1896. Photograph courtesy of Smithsonian Institution.

p. 6. First streamgaging camp, Embudo, N. Mex., 1888. USGS.

p. 8. A scientist measures streamflow near Clayton, N.J. USGS.

p. 9. Oblique aerial view of Mount St. Helens, Wash., before the May 18, 1980, eruption, as seen from Bear Cove, Spirit Lake. Photograph courtesy of USDA, Forest Service.

p. 9. Mount St. Helens in eruption, May 18, 1980. D.A. Swanson, USGS.

p. 9. Mount St. Helens reflected in Spirit Lake, 2 years after the eruption of May 18, 1980. USGS.

p. 9. San Francisco, Calif., earthquake of April 18, 1906. Hibernia Bank Building. 1906. USGS.

p. 10. Whooping crane egg. Nelson Beyer, USGS.

p. 10. A USGS scientist uses a puppet to teach a 2-week-old whooper how to probe in the wet ground, to help him learn to hunt and forage. Barbara Niccolai, USGS.

p. 11. Whooping cranes at Aransas National Wildlife Refuge. Photograph courtesy of U.S. Fish and Wildlife Service.

p. 11. Gray wolf. David Mech, USGS.

p. 11. Patuxent Wildlife Research Center, Md. USGS.

p. 12. Denver Federal Center, Colo. USGS.

p. 12. Yellow warbler. Chandler Robbins, USGS.

p. 12. Mallard ducks. Milton Friend, USGS.

p. 12. Wildlife at the Salton Sea. Milton Friend, USGS.

p. 13. Pumping ground water to support the San Joaquin Valley agricultural region. Photograph courtesy of California Department of Water Resources.

p. 14. A USGS librarian holds an extremely rare 19th century book on gems and precious stones from India. USGS.

p. 14. Sawgrass marsh. USGS.

p. 14. Mangrove. USGS.

p. 14. Salt marsh. USGS.

p. 15. Scientists use an airboat to conduct research in a wetland. USGS.

p. 15. Sea-floor map of Alderdice Bank, northwestern Gulf of Mexico. USGS.

p. 15. Menlo Park Center, Calif. Leslie Gordon, USGS.

p. 16. Sea-floor map of East Flower Garden Bank, northwestern Gulf of Mexico. USGS.

p. 16. Sea-floor map of Bouma, Rezak, Sidner, and McGrail Banks, northwestern Gulf of Mexico. USGS.

p. 17. Gene and Carolyn Shoemaker at Palomar telescope. USGS.

p. 17. An astronaut in the USGS geology training program. USGS.

p. 17. The "11-Meter" crater (left) and the Philby "A" crater (right). Both are part of the Wabar meteorite impact craters complex in the core of the Ar-Rub' Al-Khali (Empty Quarter) region of Saudi Arabia. Jeff Wynn, USGS.

p. 18. Aerial photograph of Cambridge, Mass., May 7, 1996. USGS.

p. 18. Aerial photograph of area 4 km east of Grand Coulee, Wash., July 5, 1991. USGS.

p. 18. Alaska earthquake of March 27, 1964. Fourth Avenue near C Street in Anchorage collapsed because of a landslide caused by the earthquake. Before the shock, the sidewalk on the left, which is in the graben, was at street level on the right. The graben subsided 11 ft in response to 14 ft of horizontal movement. Photograph courtesy of U.S. Army.

p. 19. Land-cover map of the 48 States. USGS.

p. 19. Water fountain. © Jupiter Images, Clipart.com.

p. 19. Alaskan coastal brown bear (female with yearlings). Steven Partridge, USGS.

p. 20. Satellite image of the Chesapeake Bay watershed. USGS.

p. 20. La Conchita, Calif.—a small seaside community along Highway 101 south of Santa Barbara. This landslide and debris flow occurred in the spring of 1995. R.L. Schuster, USGS.

p. 21. Landsat image of San Francisco Bay. USGS.

p. 21. Aerial view of the Platte River. Michael Collier, USGS.

p. 21. National Wildlife Health Center, Madison, Wis. USGS.

p. 22. Spotted salamander. USGS.

p. 22. Wood frog. USGS.

p. 22. John Wesley Powell Building, USGS National Headquarters, Reston, Va. Dave Usher, USGS.

p. 23. Sea lamprey. Photograph courtesy of U.S. Fish and Wildlife Service.

p. 24. A local catfish angler (Thomas Ellis, on left) and Ed Malindzak, a master of science student at the North Carolina Cooperative Fish and Wildlife Research Unit, with a flathead catfish captured from Contentnea Creek, N.C. Flathead catfish are not native to the Atlantic slope, and the N.C. Coop Unit is studying their effects on native riverine fishes. Photograph by Bill Pine.

p. 24. Leafy spurge, native to Eurasia, is one of the most serious weeds in the northern United States, causing millions of dollars of crop losses and control costs. Photograph by George Markham, courtesy of USDA, Forest Service.

p. 24. Yellow starthistle, a native of the Mediterranean region, is crowding out cattle forage as well as rare species such as this mariposa lily in Hells Canyon, Idaho. Photograph by J. Asher, courtesy of Bureau of Land Management.

p. 25. Spanning the southern tip of the Florida Peninsula and most of Florida Bay, Everglades National Park is the only subtropical preserve in North America. USGS.

p. 25. Center-pivot irrigation systems created these circular patterns in crop land near Garden City, Kans. The red circles indicate irrigated crops of healthy vegetation. The light-colored circles represent harvested crops. USGS.

p. 25. The Lambert Glacier in Antarctica is the world's largest glacier. An icefall feeds into the glacier from the vast ice sheet covering the polar plateau. USGS.

p. 25. Mount St. Helens in eruption on May 18, 1980; the violence of the eruption contrasts with the quiet countryside. Mount Adams in background. Skamania County, Wash. USGS.

p. 26. A scientist makes a discharge measurement, Redwell Basin, Colo. Philip Verplanck, USGS.

p. 26. A USGS scientist inspects a radio-reporting flood-warning gage at the East branch of the Rahway River, at Millburn, N.J. Dave Usher, USGS.

p. 27. A satellite dish at the EROS Data Center, Sioux Falls, S. Dak. USGS.

p. 27. Two screen captures from searches of the Geographic Names Database. USGS.

p. 28. Freeway collapse resulting from the 1994 earthquake in Northridge, Calif. USGS.

p. 28. Redoubt volcano on April 10, 1990. Drift glacier occupies the valley extending from the vent (at steam plume) to the valley bottom (visible in lower right). USGS.

p. 28. Seismogram. Dave Usher, USGS.

p. 29. A scientist returns an ice core to storage at the National Ice Core Laboratory. USGS.

p. 29. Loma Prieta, Calif., earthquake of October 17, 1989. Lack of adequate shear walls on the garage level exacerbated damage to this structure at the corner of Beach and Divisadero in San Francisco's Marina District. USGS.

p. 30. A screen shot from *The National Map* viewer shows Bethany Beach, Del. USGS.

p. 30. A screen shot from TerraServer showing a portion of a USGS topographic map of Washington, D.C. USGS.

p. 31. Photograph of Tucson, Ariz., from *The National Map*. USGS.

p. 31. The National Wetlands Research Center, Lafayette, La. USGS.

p. 32. Two employees use a GIS application to view several datasets of a given area. Sharon Cline, USGS.

p. 33. Damage to a parking garage as a result of the 1994 Northridge earthquake. D. Carver, USGS.

p. 33. The urban growth of three U.S. cities can be seen in these sets of illustrations (from USGS Circular 1252). USGS.

p. 34. Flagstaff Science Center, Ariz. USGS.

p. 34. Satellite images of Las Vegas from 1972, 1986, and 1992. USGS.

p. 34. A memorial to the 57 people who perished in the eruption of Mount St. Helens is 5 miles north of the volcano on Johnston Ridge. The ridge was named for USGS geologist David Johnston, who was killed at a nearby observation post. Photograph by Tom Iraci, courtesy of USDA, Forest Service.

p. 35. Flowering *Tamarix ramosissima* photographed at Dos Palmas Oasis (near Palm Springs) in Riverside, County, Calif. Jeffrey E Lovich, USGS.

p. 35. Zebra mussels. USGS.

p. 36. Brown tree snake. USGS.

p. 36. Multispectral false color image of Hurricane Floyd, September 15, 1999. Image courtesy of NOAA.

p. 37. Cape Hatteras National Park, N.C. Photograph courtesy of National Park Service.

p. 37. Badlands National Park, S. Dak. Photograph courtesy of National Park Service.

p. 37. Denali National Park, Alaska. Photograph courtesy of National Park Service.

p. 37. Robert H. Meade inserts a smoked glass slide into a bathythermograph . This instrument, familiar to all ocean researchers, records water temperature at various depths on the smoked glass slide. USGS.

p. 38. USGS conducts studies of dust particles in the aftermath of the collapse of the World Trade Center. Gregg Swayze and Todd Hoefen, USGS.

p. 38. Albert White Hat, Sr., explains the relationship between Sinte Gleska University and the USGS during the signing of a Memorandum of Understanding. USGS.

p. 39. Tribal personnel are trained by USGS hydrologists as part of the Bureau of Indian Affairs Water Technician Training Program, held at Central Washington University, Ellensburg, Wash. Photograph by David E. Click.

p. 39. Sampling on the La Jolla Indian Reservation, Calif., on National Water Monitoring Day 2002. Julia A. Huff and Laurel L. Rogers, USGS.

p. 40. Seismometers like this one near Mount Spurr volcano (on skyline in background) provide the Alaska Volcano Observatory with a continuous, radio-telemetered record of volcanic earthquakes. These data are used to monitor activity at the volcano and are critical to the ability of the observatory to issue timely warnings of eruptions. C. Neal, USGS.

p. 40. View from space of the Earth's Western Hemisphere. Photograph courtesy of NASA.

p. 40. View from space of the Earth's Eastern Hemisphere. Photograph courtesy of NASA.

p. 40. Seismic hazard maps are based on information about the rate at which earthquakes occur in different areas and on how far shaking extends from the earthquake source. This national map of earthquake shaking hazards shows areas vulnerable to earthquake shaking and provides information for creating and updating the seismic design provisions of building codes used in the United States.

p. 41. An 18-km-high volcanic plume from one of a series of explosive eruptions of Mount Pinatubo beginning on June 12, 1991, viewed from Clark Air Base, Philippines (about 20 km east of the volcano). David H. Harlow, USGS.

p. 41. Hurricane Mitch. Photograph courtesy of NOAA.

p. 41. Recovery of a tripod with sediment transport instrumentation used in the Boston Harbor study. USGS.

p. 42. USGS scientist Ross Stein in the Aya Sofia in Istanbul, Turkey. This ancient church is an important structure in the study of historical earthquakes. Serkan Bozkurt, USGS.

p. 43. Decisionmakers and citizens across the United States value USGS science when addressing environmental, natural-resource, and hazards-related issues. © Corbis Images.

p. 43. The *Culex pipiens quinquefasciatus* mosquito is responsible for the transmission of West Nile virus. Photograph by Stephen Higgs, Department of Pathology, University of Texas Medical Branch, Galveston, Tex.

p. 44. A USGS scientist discusses hydrologic issues with a group of future scientists. USGS.

p. 44. An elk suffers from chronic wasting disease. Photograph by Terry Kreeger, courtesy of Wyoming Game and Fish Department.

p. 45. Chemicals used every day in homes, industry and agriculture can enter the environment through our wastewater and other mechanisms. ©Hemera.com.

p. 45. A poster highlights the diversity of the USGS workforce. David Newman, USGS.

p. 46. EROS Data Center, S. Dak. USGS.

p. 46. Potomac River at low flow looking north from Chain Bridge, Washington, D.C. David Brower, USGS.

p. 46. Potomac River during flood conditions looking north from Chain Bridge, Washington, D.C. David Brower, USGS.

p. 47. In response to the growing need for information for sustainable development, the USGS conducts cooperative international projects to assess mineral, energy, water, and other natural resources throughout the world and is actively involved in research to identify new resources for the future. Jean Noe Weaver, USGS.

p. 47. Sampling freshly exposed mine waste at the Elizabeth Mine Superfund Site, Vt. USGS.

p. 47. Personal Digital Assistants are increasingly used for field data collection activities. Maura Harrison, USGS.

p. 48. Gas hydrates are a potentially huge energy resource. USGS.

p. 48. Ground-penetrating radar being tested as a future noncontact streamgaging tool on the Cowlitz River, Wash. USGS.

p. 49. New advances in genetics and molecular biology could help bring animal and plant species back from the brink of extinction. Maura Harrison, USGS.

p. 49. Keeping USGS information secure and accessible is a critical data-management issue. Maura Harrison, USGS.

p. 50. ShakeMaps did not exist in 1994 when the magnitude 6.7 Northridge, Calif., earthquake occurred. Had a ShakeMap been available for that earthquake, it could have been used to rapidly guide emergency-response teams to areas that potentially had the greatest need. This ShakeMap is made with data recorded from the Northridge earthquake. It shows that the greatest shaking occurred to the north of the epicenter and in other isolated areas, a pattern that could not have been anticipated with only magnitude and location information for the earthquake. USGS.

p. 50. A ShakeMap for a possible magnitude 7.2 earthquake near Seattle, Wash., shows estimated intensities and ground motions for a planning scenario. Such scenarios are useful for emergency response exercises and planning, as well as for understanding the potential consequences of future large earthquakes. USGS.

p. 51. This Landsat 7 image of the Mackenzie Delta, Northwest Territories, was acquired on August 28, 2002. The image is representative of the satellite data provided by the EROS Data Center to scientists, resource managers, and educators around the world. USGS.

p. 51. Washington, D.C., is visible as light blue in this EO-1 Hyperion image. USGS.

p. 51. A lightning strike on June 16, 2004, started the Pingo wildfire amidst Alaska's worst heat wave in 50 years. As of July 7, 2004, the fire had burned 180,000 acres and continued to threaten the village of Venetie, visible in the lower right corner of this Landsat image. USGS.

p. 51. When the creativity of USGS scientists is combined with the opportunities of the private sector, the American public is well served. Maura Harrison, USGS.

p. 52. Like dark fingers, cold ocean waters reach deeply into the mountainous coastline of northern Norway, defining the fjords for which the country is famous. Flanked by snow-capped peaks, some of these ice-sculpted fjords are hundreds of meters deep. USGS.

p. 52. The Richat Structure is a geological formation in the Maur Adrar Desert in Mauritania. Although it resembles an impact crater, the Richat Structure formed when a volcanic dome hardened and gradually eroded, exposing the onionlike layers of rock. USGS.

p. 52. The West Fjords are a series of peninsulas in northwestern Iceland. They represent less than one-eighth the country's land area, but their jagged perimeter accounts for more than half of Iceland's total coastline. USGS.

p. 52. Landsat data were 'draped' over data from the Shuttle Radar Topography Mission (SRTM) to create a perspective view of the Santa Barbara, Calif., region. USGS.

p. 53. The USGS map "A Tapestry of Time and Terrain" combines a shaded-relief map and a geologic map to provide a unique view of the conterminous States. USGS.

Acknowledgments

We thank the more than 100 people who provided information and images and who reviewed this Circular; we are grateful for your support and help. Above all, we thank all the USGS employees throughout our history who have contributed their energy, intelligence, and creativity to provide 125 years of relevant, impartial science for America and for the world.

Compilation:
  Kathleen K. Gohn
Layout:
  Maura H. Harrison
Editing:
  Dona Brizzi
125th Anniversary Team:
  Carolyn C. Bell
  Jon C. Campbell
  Athena P. Clark
  Joye L. Durant
  Gaye S. Farris
  Marion M. Fisher
  Leslie C. Gordon
  Daniel L. James
  Martha D. Kiger
  Tania M. Larson
  Glenn G. Patterson
  Frances W. Pierce
  Elizabeth A. Stettner
  Stephen J. Vandas
  Karen R. Wood